I0060881

# Frontiers in Clinical Drug Research-Central Nervous System
## *(Volume 1)*

### Editor

## Atta-ur-Rahman, *FRS*

*Kings College*
*University of Cambridge*
*Cambridge*
*UK*

# CONTENTS

# FOREWORD

*Frontiers in Clinical Drug Research – Central Nervous System* covers important advances in the field of drug development with particular focus on research on drugs in advanced stages of development, clinical trials and applications. This Volume published under the editorship of Prof. Atta-ur-Rahman aims at providing readers with an update on contemporary research. The research articles are comprehensive and have been written by an impressive group of authors, most of whom are well known for their own contributions in the field of drug research.

This book should be of significant interest to Ph.D. students and established researchers. I commend the editor, Prof. Atta-ur-Rahman, and all the contributors for this well written volume which highlights the quest for a cure for Neurological disease and related disorders.

*Ferid Murad,*
(Nobel Prize Physiology and Medicine 1998)
George Washington University Washington,
DC, USA

# PREFACE

*Frontiers in Clinical Drug Research – Central Nervous System* presents the latest researches on the Central Nervous System (CNS). The contributions by leading researchers in the field shed light on the development and pathophysiology of the brain and spinal cord, physiological sites of drug action in the CNS and clinical findings on drugs used to treat CNS defects due to injury or impaired development. The book also highlights other aspects of CNS medicine such as pain medicine, stem cell research, pharmacology, toxicology and translational models in animals.

In the first chapter, Odagaki has provided an update of the recent development of psychoactive agents that act as agonists or allosteric modulators at several GPCR subtypes. These are potentially useful as therapeutic drugs for mental disorders. In Chapter 2 Gosselin present the promising recent developments in the quest of modulators of glial activity, particularly emphasizing the hurdles embodied by their pharmacodynamics and pharmacokinetics specificities. Particular attention is given to neuropathic pain owing to the important literature available in the field.

Spinal Cord Injury (SCI) is a common cause of neurological morbidity and mortality, particularly affecting young adults. In Chapter 3 Fehlings et al. discuss the potential of ion-channel blockers, targeted therapeutics and monoclonal antibodies. The experiences of human SCI stem cell trials are critically appraised and a novel therapeutic target of central pattern gait generator stimulation in the process of neuro-rehabilitation is proposed. Lee and Hwang present the molecular and functional characteristics of the peptide neurotoxins targeting voltage-gated ion channels in the nervous system in chapter 4. They also discuss the current status of research and development. In the last chapter Kaur *et al.* have summarized the currently available and potential new drugs along with their mechanism of action underlying suppression of microglial activation for the treatment of White Matter Damage (WMD) in the developing brain.

I hope that the readers will find these reviews valuable and thought provoking so that they trigger further research in the quest for the development of pharmacological agents used for the treatment of allergies.

I am grateful for the timely efforts made by the editorial personnel, especially Mr. Mahmood Alam (Director Publications) and Miss Maria Baig at Bentham Science Publishers.

*Atta-ur-Rahman, FRS*
Honorary Life Fellow
Kings College
University of Cambridge
Cambridge
UK

# List of Contributors

**Charanjit Kaur**

*Department of Anatomy, MD10, 4 Medical Drive, Yong Loo Lin School of Medicine, National University of Singapore, Singapore*

**Eng-Ang Ling**

*Department of Anatomy, MD10, 4 Medical Drive, Yong Loo Lin School of Medicine, National University of Singapore, Singapore*

**Gurugirijha Rathnasamy**

*Department of Anatomy, MD10, 4 Medical Drive, Yong Loo Lin School of Medicine, National University of Singapore, Singapore*

**Jared Wilcox**

*Divisions of Orthopaedic Surgery and Neurosurgery, University of Toronto, Canada and the Division of Genetics and Development, Toronto Western Research Institute, Toronto, Canada*

**Madhuvika Murugan**

*Department of Anatomy, MD10, 4 Medical Drive, Yong Loo Lin School of Medicine, National University of Singapore, Singapore*

**Michael G. Fehlings**

*Divisions of Orthopaedic Surgery and Neurosurgery, University of Toronto, Canada and the Division of Genetics and Development, Toronto Western Research Institute, Toronto, Canada*

**Romain-Daniel Gosselin**

*Pain Centre, Department of Anesthesiology, University Hospital Centre (CHUV) and Department of Fundamental Neuroscience, University of Lausanne, Lausanne, Switzerland*

**Seungkyu Lee**

*F.M. Kirby Neurobiology Center, Children's Hospital Boston, Boston, MA, USA; Department of Neurobiology, Harvard Medical School, Boston, MA, USA*

**Simon Harris**

*Divisions of Orthopaedic Surgery and Neurosurgery, University of Toronto, Canada and the Division of Genetics and Development, Toronto Western Research Institute, Toronto, Canada*

**Sun W. Hwang**

*Department of Biomedical Sciences, Korea University College of Medicine, Seoul, Korea*

**Yuji Odagaki**

*Department of Psychiatry, Faculty of Medicine, Saitama Medical University, Saitama, Japan*

# Frontiers in Clinical Drug Research-Central Nervous System

# *(Volume 1)*

# CHAPTER 1

# Agonists and Allosteric Modulators of G Protein-Coupled Receptors as Promising Psychotropic Drugs

## Yuji Odagaki[*]

*Department of Psychiatry, Faculty of Medicine, Saitama Medical University, Saitama, Japan*

**Abstract:** During the last half a century, a lot of psychoactive agents have been developed and utilized as therapeutic drugs for mental disorders such as schizophrenia, mood disorders, anxiety disorders, and dementia. Based on the major target illnesses or symptoms, they have been classified into antipsychotics, antidepressants, mood stabilizers, anxiolytics, hypnotics, and nootropics. From a pharmacological point of view, it is well established that most of these psychotropic drugs alter neural transmission *via* classical neurotransmitters, *e.g.*, dopamine, serotonin (5-HT), norepinephrine (NE), glutamate, $\gamma$-aminobutyric acid (GABA), and acetylcholine, all of which are implicated in the maintenance and control of higher brain function and human behavior. One major molecular target of these psychotropic drugs is a G protein-coupled receptor (GPCR), at which the neurotransmitter is specifically bound. In most cases, the psychotropic drugs behave as antagonists at the GPCR. For instance, all classical antipsychotics are antagonists of dopamine $D_2$ receptors. The recent approval and great success of aripiprazole (a partial dopamine $D_2$ agonist) as an effective antipsychotic drug in many countries, has paved the way for the concept that some GPCR agonists have the potential to treat psychiatric illnesses. Interestingly, the prototypic atypical antipsychotic clozapine (or its active metabolite *N*-desmethylclozapine) behaves as an agonist at several GPCRs. It is also well known that 5-$HT_{1A}$ receptor agonists, such as buspirone and tandospirone, are efficacious anxiolytics. Another major progress in psychopharmacology in recent years is the recognition of multiple allosteric sites for many GPCRs, and many novel allosteric modulators of GPCRs have been synthesized. Though still preliminary, many studies have indicated that these allosteric modulators are promising as novel effective psychotropic drugs. In this chapter, the author will provide an update of the recent development of psychoactive agents that act as agonists or allosteric modulators at several GPCR subtypes, which are potentially useful as therapeutic drugs for mental disorders.

**Keywords:** Antipsychotic, antidepressant, anxiolytic, G protein-coupled receptor (GPCR), agonist, intrinsic activity, antagonist, allosteric modulator, dopamine, serotonin (5-HT), acetylcholine, glutamate.

*Address correspondence to Yuji Odagaki: Department of Psychiatry, Faculty of Medicine, Saitama Medical University, Moroyama-machi, Iruma-gun, Saitama, Japan; Tel: +81-49-276-1214; Fax: +81-49-276-1622; E-mail: odagaki@saitama-med.ac.jp

**Atta-ur-Rahman (Ed)**
**All rights reserved-© 2013 Bentham Science Publishers**

## INTRODUCTION

After a series of serendipitous discoveries of psychoactive drugs several decades ago, pharmaceutical companies have developed many newer therapeutic agents for the treatment of mental disorders. These new generation psychotropic drugs are generally more tolerable than their ancestors regarding the profile of adverse effects. From the viewpoint of efficacy for the targeted psychiatric symptoms, however, relatively little progress has been made. It stands to reason as long as the principle modes of pharmacological mechanisms of the newer drugs are essentially same as those of the prototypic agents.

To make a breakthrough in the development of psychotropic drugs toward the ideal ones with more efficacy and minimal adverse events, alternative mechanisms of action derived from novel pathophysiological hypotheses of mental disorders should be considered. However, it may also be true that efficacious psychotropic drugs cannot be developed irrelevant to the classical hypothesized neural transmissions implicated in pathophysiology of mental disorders. For example, despite the promising preclinical reports and expectations from clinicians, the results of clinical studies on the effects of sigma receptor ligands for mental disorders have been still controversial [1-3]. In addition, a study using single photon emission computed tomography (SPECT) with [$^{123}$I]iodobenzamide demonstrated that the possible antipsychotic activity of EMD 57445 (panamesine), one of the candidates of high affinity sigma ligands investigated, was not necessarily attributable to its affinity for sigma receptors, but simply due to the potent antidopaminergic effects of its active metabolite EMD 59983 [4].

Even if we are still stand on the so-called monoamine hypotheses of mental illnesses, it is possible to develop new therapeutic drugs with distinct mechanisms of action from those of conventional psychotropic agents, *i.e.*, dopamine $D_2$ receptor antagonism of antipsychotics, and norepinephrine (NE) and/or serotonin (5-HT) re-uptake inhibition of most antidepressants. One promising possibility is to utilize an agonist, especially a partial agonist, at several neurotransmitter receptors, and this strategy has been successfully established in some cases. An alternative gateway to the better psychotropic drugs with novel modes of action is to develop allosteric modulators of neurotransmitter receptor subtypes. In the present chapter, the author focuses on the present status and future prospects of newer psychotropic drugs and their candidates, which behave as an agonist or an allosteric modulator of G protein-coupled receptors (GPCRs).

## PHARMACOLOGICAL BACKGROUND

### Agonists and Partial Agonists

GPCRs serve as molecular targets of many psychotropic drugs. The pharmacological activity of a drug can be defined by the dual properties of affinity and intrinsic efficacy with regard to their interaction with receptors. Drugs that shift the distribution of conformations toward activation of G proteins and subsequent intracellular signaling cascades are designated as agonists. A full agonist will bind to the receptor to propagate the neurotransmission to the same extent as an endogenous ligand such as a neurotransmitter. An antagonist is defined as a ligand with affinity but without any efficacy, and thus its binding to the receptor will block the action of the endogenous neurotransmitter and an exogenous agonist.

Intrinsic activity of a ligand is defined empirically as the maximal response to the ligand as a fraction of that to a full agonist at the receptor. Partial agonists have an intrinsic activity between 0 and 1, and when they bind to the receptor they do not elicit the same conformational shift toward G protein activation and consequent biological responses as full agonists (Fig. **1**). As a consequence, they can act either as functional agonists or functional antagonists, depending on the surrounding levels of the endogenous neurotransmitters. Even when all receptors are occupied with, and activated by, high concentrations of partial agonists, there exist ceiling effects. Thus, partial agonists appear to have generally milder effects than full agonists. Furthermore, the use of partial agonists often avoids the development of adverse effects (*e.g.*, desensitization, adaptation, tolerance and dependence) that are usually associated with overstimulation of the receptors by full agonists [5, 6].

The drugs that behave as partial agonists at GPCRs have been used successfully in some cardiological applications [6]. For instance, some of β-blockers widely used for hypertensive patients (*e.g.*, pindolol), behave as a partial agonist at β-adrenergic receptors, and possess intrinsic sympathomimetic activity [7]. Another well-known example of GPCR agonists used in clinical practice is pharmacotherapy of

Parkinson's disease with dopamine agonists such as bromocriptine, pergolide, cabergoline, talipexole, pramipexole, and ropinirole [8]. Also, morphine and some synthetic opioids, which are used as effective analgesics, have agonistic properties at opioid receptors [9]. Partial agonists at $\mu$-opioid receptor, such as methadone [10] and buprenorphine [11], have been used in the substitution therapy of opioid addicts.

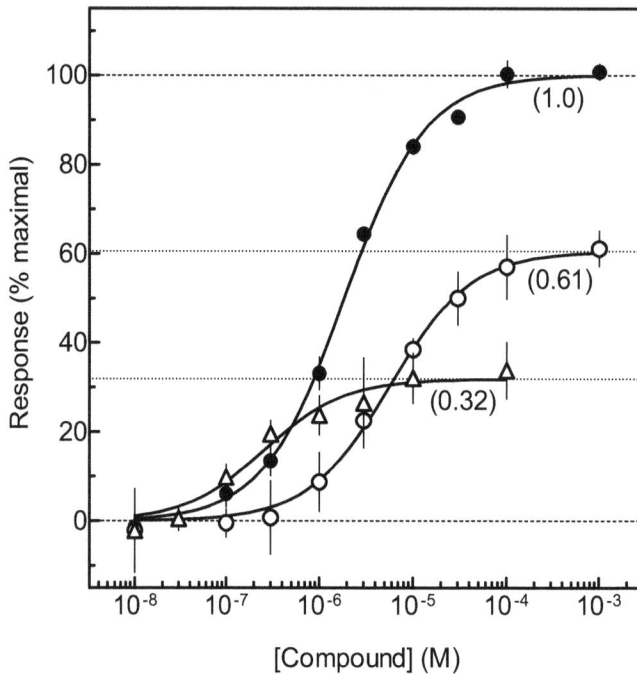

**Figure 1:** Example of concentration-response curves for the agonists with different maximal agonist effects. The responses elicited by the full agonist (•) and the two partial agonists (○, ▵) at the indicated concentrations are expressed as normalized values of the maximal response of the full agonist. The intrinsic activities [1.0 for the full agonist (•), 0.61 for the partial agonist (○), and 0.32 for the partial agonist (▵)] are indicated in parentheses.

In psychiatric pharmacotherapy, the 5-HT$_{1A}$ receptor agonist buspirone was a front runner of drugs with agonist properties at GPCRs that were successfully introduced into the treatment of mental disorders [12, 13]. Nevertheless, the great success of aripiprazole, a dopamine D$_2$ receptor partial agonist, in the treatment of schizophrenic disorders in recent several years, has brought revolution in developing of novel psychotropic drugs.

## Allosteric Modulators

The binding site of a GPCR for the endogenous ligand is termed the orthosteric binding site, which can be occupied competitively by the conventional agonists and antagonists. In addition, it has been known that multiple GPCRs have "allosteric" binding sites that are spatially and often functionally distinct from the orthosteric binding sites. To develop the compounds that act at allosteric binding sites appears an alternative and attractive approach to produce novel classes of psychotropic drugs [14-18]

Allosteric ligands can possess multiple modes of pharmacological action [19]. Some allosteric ligands are capable of receptor activation in their own right regardless of their binding to allosteric sites, but not to orthosteric binding sites, and are designated as allosteric agonists. Most allosteric modulators of GPCRs are, however, pharmacologically quiescent in the absence of an orthosteric ligand, and increase or decrease the action of an orthosteric agonist by binding at an allosteric site that leads to a change in receptor conformation (Fig. **2**). They are generally termed positive and negative allosteric modulators depending on the direction of modulation of the response elicited by an orthosteric ligand. Typically, the positive and negative allosteric modulators exert their pharmacological effects by shifting the concentration-dependent response curve to the left (Fig. **2**, right panel) and to the right, respectively. In some cases, the intrinsic efficacy of the receptor-orthosteric ligand complex can be altered. When an allosteric ligand has dual pharmacological actions, *i.e.*, stimulatory effect as an allosteric agonist and potentiating effects on the response elicited by an orthosteric ligand, it is referred to as "ago-allosteric" modulator.

As opposed to a classical agonist, positive allosteric modulators have several major advantages [21]. Above all, it is possible with relative ease to develop truly subtype selective ligands, as the allosteric binding sites are usually less conserved as compared with the orthosteric sites. Another major advantage is that allosteric modulators have lower risk of target-mediated toxicity, due to a "ceiling effect" whereby progressively increasing doses of a positive or negative allosteric modulator beyond a certain point will fail to elicit a further pharmacological response. A third advantage for allosteric modulators is related to their ability to selectively tune response only in tissues, in which the endogenous agonist exerts its pharmacological effects. Such modulators are quiescent in the absence of endogenous orthosteric activity, and thus they can process information gained from the physiology of the system to produce optimum effect, both spatially and temporally. Furthermore, most of allosteric ligands for metabotropic glutamate (mGlu) receptors are neutral, lipophilic molecules, and thus they are able to pass freely the blood brain barrier through passive diffusion [22].

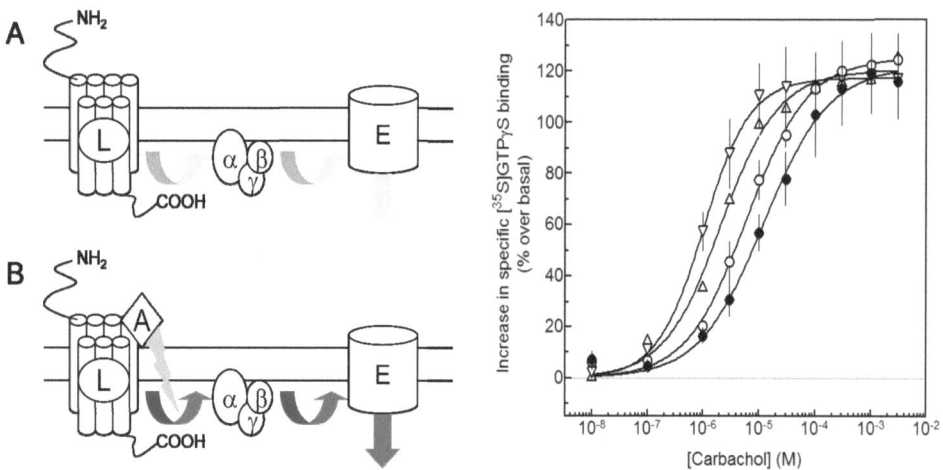

**Figure 2:** Representative allosteric modulation of the response induced by an orthosteric agonist. Left panel shows a schematic model of signaling cascade through membrane-bound GPCRs. [A] In the absence of allosteric modulator, the binding of a ligand (L) to the GPCR induces conformational change in the receptor, which then facilitates or inhibits the effector (E) function through heterotrimeric ($\alpha$, $\beta$, and $\gamma$) G proteins. [B] When an allosteric ligand (A) binds to a topographically distinct (allosteric) site on the GPCR, the affinity and/or efficacy of the orthosteric ligand is altered. Right panel shows an example of positive allosteric modulator of $M_1$ muscarinic acetylcholine receptor (mAChR) function. The increase in specific $[^{35}S]GTP\gamma S$ binding to $G\alpha_q$ elicited by carbachol was determined by the antibody-capture scintillation proximity assay/$[^{35}S]GTP\gamma S$ binding using the anti-$G\alpha_q$ antibody in rat hippocampal membranes, in the absence ($\bullet$), and presence of 1 $\mu M$ ($\circ$), 10 $\mu M$ ($\triangle$), and 100 $\mu M$ ($\triangledown$) VU 0029767 [20].

For clinical use, only two allosteric modulators have been marketed to date [18]. Cinacalcet, a positive allosteric modulator of the calcium sensing receptor, is used for the treatment of secondary hyperparathyroidism and parathyroid carcinoma. Another one is maraviroc (UK-427, 857), a noncompetitive allosteric antagonist of the chemokine receptor, CCR5, which has been approved for the treatment of HIV infection.

Unfortunately, no psychotropic drugs have been clinically available yet, whose principal mechanisms of action are located on allosteric sites, but not on orthosteric sites, of GPCRs. Although it has been shown that *N*-desmethylclozapine, an active metabolite of the atypical antipsychotic clozapine, potentiates *N*-methyl-D-aspartate (NMDA) receptor activity probably through the allosteric interaction with the $M_1$ muscarinic acetylcholine receptor (mAChR) [23], this might be only an ancillary pharmacological property of clozapine, one of the most dirty drugs targeting multiple receptor subtypes. Nevertheless, development of the compounds acting at allosteric sites of GPCRs may pave the way for novel epoch-making psychotropic drugs that are able to stabilize disorganized higher brain functions in the patients suffering from mental disorders. In the present chapter, theoretical background and promising compounds in the near future have been concisely reviewed as to the allosteric modulators for several subtypes of mAChRs and mGlu receptors.

## DRUGS THAT BEHAVE AS AN AGONIAT AT GPCRs

### 5-HT$_{1A}$ Receptor

The 5-HT receptor family is composed of at least 14 distinct subtypes, which are subclassified into 5-HT$_1$ to 5-HT$_7$ receptors. The 5-HT$_{1A}$ receptor, a member of 5-HT$_1$ receptors, has been of particular interest because its function is involved in many physiological phenomena including control of mood, impulsivity, cognition, and memory.

Buspirone has been widely used for the treatment of anxiety or dysphoria of moderate intensity since 1980s [12, 13, 24]. This compound belongs chemically to the azapirone derivatives, and behaves as a partial agonist at 5-HT$_{1A}$ receptor, without any affinity for the benzodiazepine receptors. As a consequence, it does not exert the adverse events related to the benzodiazepine receptors, such as sedation, hypnotic effects, muscle relaxation, tolerance, and withdrawal syndrome. Although the mechanisms of action of buspirone are incompletely understood, it has been hypothesized that the anxiolytic effect is mediated through a reduction of the firing rate of dorsal raphe serotonergic neurons and decreased synthesis and release of 5-HT, resulted from its stimulatory effects on somatodendritic 5-HT$_{1A}$ autoreceptors [12, 24].

In Japan, buspirone has not been marketed, and instead of buspirone, another azapirone derivative tandospirone (SM   3997) is approved as an anxiolytic drug [25]. Tandospirone is also a partial agonist at 5-HT$_{1A}$ receptor subtype, with an intrinsic activity at least comparable to buspirone [26-28].

In addition to the standardized anxiolytic effects in the patients with generalized anxiety disorder [29, 30], marginal to substantial antidepressant-like effects have been reported as to the azapirones with 5-HT$_{1A}$ receptor agonistic properties, such as buspirone [31], gepirone [32-34], and ipsapirone [35, 36]. In line with the potential antidepressant effects of these 5-HT$_{1A}$ receptor partial agonists, diverse preclinical and clinical studies have indicated that dysfunctional states of 5-HT$_{1A}$ receptors are implicated in the pathogenesis of major depressive disorders [37].

5-HT$_{1A}$ receptors are located presynaptically on 5-HT cell bodies in the dorsal and medial raphe nuclei, as well as postsynaptically in various brain regions including limbic systems. In contrast to their anxiolytic effects, it is suggested that the antidepressant-like effects of 5-HT$_{1A}$ receptor agonists are mediated by postsynaptic 5-HT$_{1A}$ receptors [38-40]. The therapeutic effects of 5-HT$_{1A}$ receptor agonists as antidepressants require repeated administration. This delay can be explained by the time-lag necessary for adaptation of 5-HT neurons subsequent to subsensitivity of presynaptic 5-HT$_{1A}$ receptors in raphe nuclei induced by long-term 5-HT$_{1A}$ receptor stimulation [40, 41].

As antidepressants, the clinical efficacy of the 5-HT$_{1A}$ receptor partial agonists is, at least when used as monotherapy, often limited or insufficient [35, 36, 42], which makes most clinicians disenchanted [41]. One of the possible theoretical explanations for the limited efficacy and poor tolerability of these 5-HT$_{1A}$ receptor partial agonists has been raised from a pharmacokinetic point of view [41]. In order to avoid such undesirable pharmacokinetic properties of the azapirones, *i.e.*, marked variations in plasma concentrations derived from their rapid absorption and short elimination half-life, the extended-release formulation of gepirone has been used in recent clinical trials [31, 32, 34].

Another important issue to be considered according to the clinical antidepressant efficacy of the 5-HT$_{1A}$ receptor agonists is whether optimal intrinsic activity can be determined [40]. Preclinical studies clearly indicated that the intrinsic activity of the 5-HT$_{1A}$ receptor agonists were correlated with the antidepressant effects predicted by means of forced swimming test [43]. Then, 5-HT$_{1A}$ receptor full agonists might be superior to partial agonists as effective antidepressants. It is, however, uncertain whether the data derived from preclinical studies are applicable directly to clinical situations. Moreover, even if full agonists are preferable as antidepressants, the optimal intrinsic activity for anxiolytic efficacy might be unequal to that. In fact, it has been shown that 5-HT$_{1A}$ receptor agonists are not anxiolytic but even anxiogenic in some experimental paradigms [44]. Anxiety is closely coherent to depression, and often appears as co-morbidity of depressive disorders. The complicated situations in clinical practice should be taken into consideration in developing new antidepressant/anxiolytic drugs with 5-HT$_{1A}$ receptor agonist properties.

Interestingly, besides azapirone class anxiolytic drugs such as buspirone, several clinically used compounds behave as 5-HT$_{1A}$ receptor agonists. For example, an atypical antidepressant trazodone, along with its active metabolite *m*-cholophenylpiperazine (*m*-CPP), is a 5-HT$_{1A}$ receptor partial agonist [28]. Also, one constituent of herbal extracts of yokukansan, a traditional medicine used for the management of behavioral and psychological symptoms of dementia (*e.g.*, aggression, anxiety, agitation, and hallucinations), was shown to behave as a 5-HT$_{1A}$ receptor partial agonist [45].

Another major issue is the possible importance of 5-HT$_{1A}$ receptor functionality in pathophysiology of schizophrenia and potential benefit of 5-HT$_{1A}$ agonism in mechanisms of action of antipsychotic drugs [46, 47]. Several lines of investigations including human postmortem studies have indicated elevated 5-HT$_{1A}$ receptors in the brain, especially in frontal cortical areas, of patients suffering from schizophrenic disorders. Although the pathophysiological significance of this finding remains to be elucidated, the importance of 5-HT$_{1A}$ receptors and their mediated signaling pathways as a promising target of newer antipsychotic drugs has been a focus of interest of recent publications [48, 49]. The prototypic atypical antipsychotic clozapine has been shown to be a partial agonist at 5-HT$_{1A}$ receptors [50]. This pharmacological profile is shared by several clinically available antipsychotics (mostly atypical, though not exclusively), such as aripiprazole, ziprasidone, quetiapine, perospirone, lurasidone, nemonapride, and asenapine (ORG 5222), as well as putative atypical antipsychotics in the investigational stage [51-54]. The atypical antipsychotics are superior to the typical drugs in efficacy for improving negative symptoms of schizophrenia, and 5-HT$_{1A}$ receptor partial agonist properties may be contributing to this superiority [49]. Furthermore, the cognitive enhancing properties of 5-HT$_{1A}$ receptor agonists have been widely documented [48, 55]. Since the cognitive impairments associated with schizophrenia are of key importance for work and social function [56], further preclinical as well as clinical investigations are necessary as to the relevance of 5-HT$_{1A}$ receptor stimulation in schizophrenic patients.

## 5-HT$_{2C}$ Receptor

The 5-HT$_{2C}$ receptor is also implicated in multiple physiological functions related to food intake, reproductive behavior, mood, and anxiety. In psychiatric field, this receptor subtype has been generally assumed to be involved in weight gain and related metabolic syndrome, burdensome adverse effects associated with some atypical antipsychotic drugs like clozapine and olanzapine ([57], but see also [58]). Recent preclinical studies have indicated that the 5-HT$_{2C}$ receptor agonist WAY 163909 [59] possesses therapeutic potential as an atypical antipsychotic [60] and as an antidepressant drug [61]. Potential antidepressant-like effects of other 5-HT$_{2C}$ receptor agonists (WAY 161503, Ro 60-0175, and Ro 60-0332) were also indicated by several animal behavioral studies [62-64]. On the other hand, Dekeyne *et al.* [65] provided the data indicating antidepressant and anxiolytic properties of a novel compound S32006, a selective *antagonist* at 5-HT$_{2C}$ receptors. In this regard, it is of interest to note that agomelatine, a recently licensed antidepressant [66], is an agonist at melatonin receptors with an antagonistic property at 5-HT$_{2C}$ receptors. The relevance of 5-HT$_{2C}$ receptor antagonism in the antidepressant action of agomelatine in humans is, however, still controversial [67, 68]. The 5-HT$_{2C}$ story in antidepressant efficacy is further complicated with the findings that agomelatine is a neutral antagonist whereas S32006 behaves as an *inverse agonist* in constitutively active situations [69].

It is likely that 5-HT$_{2C}$ receptor is a promising target in the development of new psychotropic drugs. Nevertheless, further investigations are necessary to clarify the importance of 5-HT$_{2C}$ receptors in brain function and behavioral control, and to determine the potentiality of modulation of 5-HT$_{2C}$ receptor function for the treatment of mental disorders.

## 5-HT$_4$ Receptor

It has been demonstrated that 5-HT$_4$ receptor agonists are likely to improve cognitive function and memory *via* both direct and indirect mechanisms, which raises the idea that they potentially serve as adjunctive therapeutics for improving cognitive impairments in antipsychotics-treated schizophrenic patients, and as cognitive-enhancing and/or neuroprotective drugs for the neurodegenerative illnesses such as Alzheimer's disease [70, 71]. Also, it has been suggested that 5-HT$_4$ receptor agonists are promising as faster-acting antidepressant-like drugs, based on the results of behavioral, neurochemical, and neurophysiological animal experiments [72, 73]. It is of interest that these antidepressant-like effects of 5-HT$_4$ receptor agonists were evident after short-term (3-days) administration, indicative of potentially more rapid onset of action than the update available antidepressants. In addition to the putative antidepressant-like effects when administered alone, *in vivo* microdialysis data have shown that increase in extracellular 5-HT levels by paroxetine is augmented by 5-HT$_4$ receptor agonist [74]. There appeared no clinical studies available on the potential efficacy of 5-HT$_4$ receptor agonists.

## 5-HT$_6$ Receptor

The 5-HT$_6$ receptor is almost exclusively distributed in the central nervous system (CNS), and thus selective ligands targeting this receptor subtype are potentially beneficial psychotropic drugs with minimum peripheral side effects. Recently, several selective agonists and antagonists at 5-HT$_6$ receptor became available in preclinical studies, and their potential effectiveness as cognitive enhancers, antidepressants, and anxiolytics have been intensively explored. Carr *et al.* [75] reported that the two selective 5-HT$_6$ receptor agonists, WAY 208466 and WAY 181187, showed both antidepressant-like and anxiolytic-like effects in rat behavioral tests. In support of these results, another 5-HT$_6$ receptor agonist 2-ethyl-5-methoxy-*N,N*-dimethyltryptamine          (EMDT)          possessed antidepressant-like effects in behavioral and immunohistochemical paradigms in mice, whereas the 5-HT$_6$ receptor antagonist SB 271046 prevented the antidepressant-like properties of EMDT and fluoxetine [76]. Furthermore, it has been suggested that several 5-HT$_6$ receptor agonists may be effective for the treatment of obsessive compulsive disorder (OCD) [77], as well as cognitive impairment [78, 79]. Paradoxically, there have been many preclinical studies indicating that 5-HT$_6$ receptor *antagonists* are also effective as potential antidepressants, anxiolytics, and cognitive enhancers [80-82]. The interpretation of the relevance of 5-HT$_6$ receptor in psychiatric pharmacotherapy is further complicated by the findings that many, but not all, typical and atypical antipsychotics as well as antidepressants are potent antagonists at this receptor subtype [83-85]. Future studies are necessary to explore the underlying mechanisms to fully explain the apparently discrepant results for the potential therapeutic efficacy by both agonists and antagonists at 5-HT$_6$ receptor.

## Dopamine D$_1$ Receptor

Currently available antipsychotics have only limited efficacy in the treatment of negative symptoms and cognitive deficits in schizophrenic patients. Since both clusters of symptoms are supposed to arise from a dopaminergic deficit in the prefrontal cortex or hypofrontality [86], wherein dysfunction of dopamine D$_1$ receptors is highly suggested [87, 88], stimulation of the dopamine D$_1$ receptors appears a rational strategy for the better outcome and higher social functioning of the schizophrenic patients. Indeed, several preclinical studies have supported this notion. For instance, Castner *et al*. [89] revealed that co-administration of the dopamine D$_1$ receptor agonist ABT 431 (prodrug of A 86929) reversed haloperidol-induced working memory deficits in monkeys. McLean *et al*. [90] reported that SKF 38393, a partial agonist at dopamine D$_1$-like receptors, significantly ameliorated the cognitive deficits induced by sub-chronic phencyclidine (PCP) treatment in rat behavioral models. Dopamine D$_1$ receptor agonists such as A 77636 and SKF 38393 were reported to ameliorate ketamine-induced spatial working memory deficits in rhesus monkeys [91]. Salmi *et al*. [92] also pointed out the potential clinical utility of a full dopamine D$_1$ receptor agonist dihydrexidine in various CNS disorders including schizophrenia. Of particular interest is *l*-stepholidine, a tetrahydroberberine alkaloid isolated from the Chinese herb *Stephania intermedia*, since it has been shown to possess a dual action at dopamine receptors, *i.e.*, antagonistic at dopamine D$_2$ receptors and agonistic at $D_1$ receptors [93, 94]. In animal models, it shows efficacy as an atypical antipsychotic drug like clozapine [95]. As *l*-stepholidine has poor oral bioavailability limiting its practical application, its derivative bi-acetylated *l*-stepholidine has been introduced recently by Guo *et al.* [96], who showed that this compound was not only effective against the hyperactivity, but also improved the sensorimotor gating deficit, social withdrawal and cognitive impairment, in the animal models of schizophrenia.

It is also noteworthy that the prototypic atypical antipsychotic clozapine has been reported to be an agonist at dopamine $D_1$ receptors in *in vivo* thermoregulatory experiments in rats [97]. However, the author is not aware of the direct evidence indicating that either clozapine or its active metabolite *N*-desmethylclozapine has dopamine $D_1$ receptor agonist properties *in vitro*.

## Dopamine $D_2$ Receptor

The potential benefit of the dopamine $D_2$ receptor partial agonists in the treatment of schizophrenia was originally hypothesized in 1970s, based on the concept that the drugs behaving as agonists solely at autoreceptors, but not at postsynaptic dopamine receptors, serve as effective antipsychotics [98]. Despite the subsequent eager efforts with the candidate compounds such as apomorphine, bromocriptine, *n*-propyl-norapomorphine (NPA), (-)-3-(3-hydroxyphenyl)-*N*-n-propylpiperidine [(-)-3PPP] (preclamol), talipexole (B-HT 920), roxindole (EMD 49980), OPC-4392, terguride, pramipexole, and SDZ-HDC-912, none of these compounds had been successfully developed as a commercially available antipsychotic drug for a long time [99]. At last, aripiprazole was launched into market as a "dopamine system stabilizer" or "Godilocks antipsychotic" [100, 101]. Still up to the present, aripiprazole remains only one antipsychotic drug with such unique pharmacological properties [102]. In my opinion, the principal reason of this fact is derived from the lower intrinsic activity of aripiprazole compared with other dopamine $D_2$ receptor partial agonists, and this matter was discussed previously in detail [103]. The most appropriate level of blockade or activation of dopamine $D_2$ receptor-mediated signaling may be individually different from patient to patient, and dependent on pathological phase even in the same patient. In order to fine-tune the dysregulated activity into the well-stabilized range, it would be preferable that several dopamine $D_2$ receptor partial agonists with different intrinsic activities become clinically available. In this sense, bifeprunox had been a promising compound as a successor of aripiprazole, because it might have an intrinsic activity somewhat higher than aripiprazole [104-106]. In a randomized, double-blind, placebo-controlled, multi-center study, 20 mg of bifeprunox was found to be significantly more effective than placebo in reducing symptoms of schizophrenia, with a low incidence of side effects [107]. However, further development of this drug was unfortunately ceased in the end.

Clozapine is an antagonist (or inverse agonist) at dopamine $D_2$ receptors on its own, but it should be noted that its principle active metabolite *N*-desmethylclozapine behaves as a partial agonist at dopamine $D_2$ receptors [108]. This unique pharmacological profile raised a possibility that this compound served as a novel antipsychotic with pro-cognitive efficacy [109]. However, clinical data on its antipsychotic efficacy are scant up to the present, and the limited information tells us that the results on its efficacy and safety in schizophrenic patients are disappointing [109, 110].

Memantine is nowadays prescribed widely for the treatment of Alzheimer's disease, and its primary pharmacological action is commonly considered to reside in the antagonism at NMDA receptors [111]. Interestingly, it is reported that memantine also behaves as an agonist at dopamine $D_2$ receptors [112]. The relevance of these findings to its clinical benefits and/or adverse events remains unclear.

**Dopamine $D_3$ Receptor**

It was reported that dopamine $D_3$ receptor agonist, (+)-PD 128907, displayed an atypical antipsychotic-like profile, similar to clozapine in rodent models [113]. However, the selectivity of this compound to dopamine $D_3$ receptor over $D_2$ subtype is not sufficiently high, and it possesses agonistic properties at dopamine $D_2$ receptor [114], which are possibly contributing to beneficial antipsychotic-like effects as described above. Consequently, there is scant evidence indicating that dopamine $D_3$ receptor activation is useful in the treatment of schizophrenic disorders. Although it has been reported that (+)-PD 128907 protects the toxic effects of cocaine in mice, mainly *via* stimulation of dopamine $D_3$ receptors [115], the clinical relevance of these findings is unclear.

Cariprazine (RGH 188) is a potent dopamine $D_3/D_2$ receptor partial agonist (or an antagonist depending on the experimental systems [116]), with ca. 10-fold preference for the $D_3$ receptor. In the assay systems for inhibition of forskolin-stimulated cAMP accumulation in CHO cells expressing human dopamine $D_2$ and $D_3$ receptors, cariprazine showed the intrinsic activities comparable to aripiprazole [117]. In preclinical studies, cariprazine was shown to be a promising candidate as an atypical antipsychotic drug with beneficial effects on learning performance [118]. The results of clinical trials of this compound for schizophrenia and bipolar disorders will be disclosed in near future [119]. It is, however, also indefinite to what extent the agonistic effects of cariprazine on dopamine $D_3$ receptors contribute to the preclinical and clinical efficacy, since it also behaves as an agonist at dopamine $D_2$ and $5\text{-}HT_{1A}$ receptors [116], both of which are promising molecular targets of several novel antipsychotics as described above. Further well-designed clinical trials focusing on the functional role of the dopamine $D_3$ receptors in human cognitive function, and on the possible pro-cognitive properties of cariprazine in schizophrenic patients, are needed [120].

## Muscarinic Acetylcholine Receptors (mAChRs)

The mAChR family comprises five molecular distinct subtypes termed $M_1$ to $M_5$, which in the CNS play a key role in the regulation of cognition, attention, mood, and nociception, as well as other brain functions such as control of motor and vegetative systems. Several lines of evidence strongly suggest that alterations in signaling pathways mediated *via* these mAChR subtypes are implicated in Alzheimer's disease [121], and several mental disorders like schizophrenia [122]. Accordingly, these mAChR subtypes serve as rationally promising molecular targets for the development of effective drugs for the treatment of these neurodegenerative and functional neuropsychiatric disorders [123-125]. Unfortunately, all mAChR ligands used in early preclinical and clinical studies lacked true receptor subtype selectivity, probably because of the highly conserved homology sequence for the orthosteric acetylcholine binding site [126]. This is the main reason why mAChRs are collectively discussed here in the present section.

Based on the rational hypothesis, considerable preclinical and clinical research efforts have been made to develop efficacious mAChR agonists (especially functional $M_1$ agonists, see [127, 128]) for the treatment of cognitive impairments associated with Alzheimer's disease and other types of dementia. However, no or little clinically significant benefit has been reported with these compounds in early studies, and further development of most of these compounds was discontinued [128-131].

Xanomeline was originally reported to be a mAChRs agonist with preferential functional selectivity to $M_1$ subtype over other mAChRs and other neurotransmitter receptors *in vitro* and *in vivo* [132]. Later, it was often referred to as a $M_1/M_4$ preferring mAChR agonist [133]. This subtype selectivity profile has been claimed to explain its limited parasympathomimetic side effects mediated *via* peripheral $M_2$ and $M_3$ receptor subtypes. As a result, this compound attracted much attention of preclinical and clinical researchers, as a promising drug for the treatment of Alzheimer's disease [133]. In a randomized, double-blind, placebo-controlled, multicenter trial, xanomeline provided significant cognitive effect compared to placebo at the highest dose, whereas 52% of patients discontinued therapy in the high-dose arm of xanomeline-treated group because of adverse events, mainly gastrointestinal in nature [134, 135]. These findings raised the question in terms of the subtype selectivity of xanomeline, determined from earlier studies [136]. Although some selectivity for $M_1$ over $M_2$ was reported for xanomeline [124, 126, 137], much higher selectivity to $M_1$ over other mAChR subtypes should be necessary to separate the beneficial pro-cognitive effects from undesirable parasympathomimetic adverse effects.

In the above-mentioned clinical trial, of particular interest were the findings that xanomeline produced prominent dose-dependent beneficial effects on psychological and behavioral disturbances associated with Alzheimer's disease, such as vocal outbursts, suspiciousness, delusions, agitation, hallucinations, and compulsiveness [134, 135]. The favorable effectiveness of xanomeline on these psychosis-like symptoms raised the possibility that this mAChRs agonist might provide a novel approach for the treatment of schizophrenia [136, 138]. In support of this concept, several preclinical results in rodents indicate that xanomeline possesses antipsychotic-like profile, similar to atypical antipsychotics like clozapine, in electrophysiological, neurochemical, and behavioral paradigms [139-141]. In some behavioral models, xanomeline was also reported to exhibit the effects similar to those elicited by typical antipsychotic drugs like haloperidol [141-143]. The antipsychotic-like effects of xanomeline, without causing extrapyramidal symptoms, were also reported in non-human primates [144]. Following these preclinical encouraging reports, a pilot study with double-blind and placebo-controlled design was performed to examine the possible efficacy of xanomeline on clinical outcomes in subjects with schizophrenia [145]. In addition to the significant improvement in psychiatric symptoms, the subjects in the xanomeline-treated group showed significant improvements in some subsets of cognitive test battery, such as verbal learning and short-term memory. Unfortunately, however, the development of xanomeline as an antipsychotic has been discontinued [128]. Interestingly, several studies using mAChR subtype knockout mice indicate that antipsychotic-like effects of xanomeline are mediated predominantly *via* $M_4$ subtype, rather than $M_1$ receptor [146-148].

With possible preferential effects of mAChRs agonists as cognitive enhancers and/or antipsychotics in mind, it is again necessary to mention the pharmacological profiles of clozapine, and its major active metabolite *N*-desmethylclozapine. In 1990s, it was shown that clozapine itself behaved as a full or partial agonist at cloned $M_1$, $M_2$, $M_3$, and $M_4$ mAChRs expressed in CHO cells [149-151]. Subsequently, Sur *et al.* [23] reported that *N*-desmethylclozapine was a relatively potent partial agonist at $M_1$ mAChR subtype, by interacting with a site that does not fully overlap with acetylcholine orthosteric site. Whether allosteric or orthosteric, this interesting paper was followed by several studies focusing on the agonist properties of this compound at mAChRs. Weiner *et al.* [152] reported that *N*-desmethylclozapine had agonistic properties at all five mAChR subtypes, whereas clozapine itself behaved as less efficacious agonist at $M_1$, $M_2$, and $M_4$ mAChRs, but as an antagonist at $M_3$ and $M_5$ subtypes. Davies *et al.* [153] also reported the agonist effects of clozapine at $M_1$ mAChR, and those of *N*-desmethylclozapine at $M_1$ and $M_5$ mAChRs. On the other hand, Thomas *et al.* [154] demonstrated that the agonist effects of *N*-desmethylclozapine were detectable in human recombinant and rat native $M_1$ mAChRs, but not in human native $M_1$ subtype in postmortem brain tissue. Thus, the available information is still controversial. Further investigations are necessary to clarify whether and how the agonist effects of clozapine (and its metabolites) at each mAChRs, if any, contribute to the clinical unique profile of clozapine.

Extensive efforts to develop the selective orthosteric agonists at mAChR subtypes are still ongoing. For example, LY 593093 was recently reported as a selective orthosteric partial agonist at $M_1$ mAChR [155]. This compound stimulated $M_1$ mAChR-mediated β-arrestin recruitment *in vitro*. Additionally, it was also active in *in vivo* experiments, as demonstrated by the effects on phosphoinositide hydrolysis in hippocampus, as well as beneficial efficacy on spatial learning paradigm.

### Group II mGlu Receptors (mGlu$_2$ and mGlu$_3$)

Glutamatergic abnormalities have long been postulated in psychiatric disorders, in particular schizophrenia. According to the glutamate hypothesis of schizophrenia, compounds that can correct or modulate dysfunctional glutamatergic neurotransmission are supposed to be useful as effective therapeutic drugs for schizophrenia. However, the drugs targeting ionotropic glutamate receptors are not considered therapeutically useful, because of the ubiquitous involvement of these receptors in mediating fast synaptic transmission throughout the CNS. An alternative approach is to target mGlu receptors that modulate synaptic neurotransmission with heterogeneous localization and distinct functional properties. Among eight subtypes of mGlu receptors, group II mGlu receptors, comprising mGlu$_2$ and mGlu$_3$, have been of particular interest as a potential target of novel antipsychotic drugs devoid of antagonism at dopamine $D_2$ receptors [156]. Accumulating evidence shows that group II mGlu receptors agonists exhibit significant beneficial pharmacological effects in numerous experimental models of psychiatric disorders including schizophrenia [157, 158].

The possible antipsychotic-like effects of an agonist at group II mGlu receptors were first described in 1988, with regard to LY 354740 in PCP model [159]. Subsequently, several drugs with the similar pharmacological properties have been developed, and the promising antipsychotic-like properties of these drugs have been confirmed in varied preclinical studies. These include LY 379268 [160-163], MGS 0008 and MGS 0028 [164], LY 544344 (prodrug of LY 354740) [165], and LY 404039 [166]. Interestingly, some beneficial effects on cognitive impairments were also reported for some of these group II mGlu receptors agonists [159, 167], whereas this issue appears controversial [156, 158, 168]. By means of knockout mice lacking either or both of mGlu receptor subtypes, it has been speculated that the antipsychotic-like effects of group II mGlu receptors agonists demonstrated in animal models are attributable to $mGlu_2$, but not $mGlu_3$ [169, 170].

Based on the above-mentioned encouraging preclinical results, one Phase 2 clinical trial was performed with a group II mGlu receptors agonist in schizophrenic patients [171]. In this randomized, double-blind, placebo-controlled study, the effect of LY 2140023, a pro-drug of LY 404039 synthesized to overcome its low oral bioavailability, was evaluated, with olanzapine as an active control. The results showed significant improvements in both positive and negative schizophrenic symptoms in the patients treated with LY 2140023, to the same extent as in the olanzapine-treated group. There were no disagreeable adverse events, such as prolactin elevation, extrapyramidal symptoms, or weight gain, in the group treated with LY 2140023. However, subsequent another Phase 2 study failed to reveal significant efficaciousness of LY 2140023, when compared to the placebo group [172].

Even if these compounds are proven to be effective antipsychotics, the fundamental thesis concerning the roles of mGlu receptors played in the antipsychotic effects of these compounds, has been critically questioned by Seeman [173]. Thus, LY 404039 [174] and other related compounds, such as LY 354740 and LY 379268 [175], are not only group II mGlu receptors agonists, but also equipotent ligands at high-affinity states of dopamine $D_2$ receptors with partial agonist properties. It remains to be addressed whether effective antipsychotic drugs can be developed entirely irrelevant to the dopamine hypothesis. However, these findings have not been replicated by others [176, 177], and the data using knockout mice indicate that dopamine $D_2$ antagonists and group II mGlu receptors agonists exert antipsychotic effects independently, through dopamine $D_2$ receptors and $mGlu_2/mGlu_3$ receptors, respectively [176].

In healthy human subjects, LY 354740 attenuated ketamine-induced working memory impairments [178]. In this case again, the question remains to be addressed whether the observed effects of these compounds are attributable solely to their pharmacological properties as glutamatergic ligands. It has also been reported that ketamine behaves as an agonist at dopamine $D_2$ receptors on its own [179]. However, it should be mentioned at the same time that the data are inconsistent as to the possibility that psychotomimetics like PCP and ketamine directly act on dopamine $D_2$ receptors [180].

Glutamatergic system is also implicated in the pathophysiology of depression and anxiety, and several hypotheses have been raised according to the possible development of novel therapeutic drugs with primary site of action on mGlu receptors [181]. One possibility is the potential anxiolytic-like effect of group II mGlu receptors agonists [182], though several reports have suggested that group II mGlu receptors *antagonists* also show anxiolytic-like effects in preclinical models of anxiety [181]. As summarized by Palucha and Pilc [181], the possible anxiolytic-like effects of group II mGlu receptors orthosteric agonists have been evaluated mainly by utilizing LY 354740 in a variety of animal models of anxiety, with promising results in most studies. The anxiolytic potential of LY 354740 has been experimentally confirmed in some human studies using fear-potentiated startle paradigm [183], and $CO_2$-induced panic provocation [182]. In the Phase 2 study for the panic disorder patients, the anti-panic effects of LY 354740 were not indicated [184]. Its low bioavailability following oral administration provoked the pharmaceutical company to synthesize LY 544344, a prodrug of LY 354740 with improved bioavailability [165]. In healthy human volunteers, the anti-anxiety effects of LY 544344 were determined in panic symptoms experimentally induced by cholecystokinin tetrapeptide (CCK-4) [185]. The results were marginal or equivocal, because the anxiolytic effects were only significant, when the two subjects, who did not show the reduction in CCK-4-induced ACTH release by LY 544344, were removed out of twelve subjects. The randomized, double-blind, parallel, placebo-controlled clinical trial was designed to evaluate the efficacy, safety, and tolerability of LY 544344 in generalized anxiety disorder patients [186]. Unfortunately, this trial was discontinued early based on the findings of convulsions reported in preclinical studies [187]. Nevertheless, the reported results indicated that higher dose of LY 544344 was effective in improvement of anxiety symptoms as compared to placebo.

## Miscellaneous

Besides the compounds acting as agonists at GRCR subtypes described above, many GPCR agonists have been raised as novel candidate psychotropic drugs. For instance, there have been several preclinical reports indicating the possible cognitive enhancing effects of selective dopamine $D_4$ receptor agonists, such as PD 168,077 [188] and A-41299 [189]. A potential atypical antipsychotic, L 15063, has unique pharmacological properties, *i.e.*, partial dopamine $D_4$ receptor agonism, in combination with dopamine $D_2/D_3$ receptors antagonism, and 5-$HT_{1A}$ receptor agonism [190]. This compound was shown to be efficacious in animal models of cognitive deficits and negative symptoms of schizophrenia [191]. Another example is LSP 1-2111, the group III mGlu receptors orthosteric agonist, with preference to $mGlu_4$ receptor. This agonist was reported to possess anxiolytic [192] and antipsychotic [193] effects.

## DRUGS THAT BEHAVES AS AN ALLOSTERIC MODULATOR AT GPCRs

## $M_1$ mAChR

As described above, $M_1$ mAChR is one of the potential molecular targets for the treatment of several neuropsychiatric disorders, including Alzheimer's disease and schizophrenia. Considerable efforts have been centered on development of selective $M_1$ mAChR agonists. Unfortunately, these efforts have been largely unsuccessful, because of the highly conserved amino acid sequence of the orthosteric acetylcholine binding site, across five mAChR subtypes. As an alternative approach, intensive endeavors have been made, in recent two decades, to develop the compounds that can stimulate $M_1$ mAChR by acting at allosteric sites.

The first identified compound of this type was brucine, which behaved as a positive allosteric modulator of $M_1$ mAChR *in vitro* [194-197]. However, brucine elicited only a two- to three-fold increase in affinity for acetylcholine, even at the high concentrations. Therefore, brucine was unlikely suitable to be utilized as an allosteric potentiator in *in vivo* experiments. Nevertheless, the functional selectivity of brucine for $M_1$ mAChR opened up a new concept that it will be possible to develop beneficial drugs with absolute selectivity for each mAChR subtypes.

Using a high-throughput functional screening, Spalding *et al.* [198] identified a compound designated as AC-42, which showed a potent and efficacious agonist activity at $M_1$ mAChR subtype, with no detectable agonist properties at other mAChRs. The experiments with a series of chimeric receptors showed that AC-42 activated the receptor, through regions that are distinct from those used by orthosteric agonists. Thus, this compound was supposed to be a selective allosteric agonist at $M_1$ mAChR subtype, which was pharmacologically verified by Langmead *et al.* [199]. However, AC-42 and its analogues [200] had limited potency and pharmacokinetic properties unsuitable for use in tissue or animal model studies.

More recently, another selective $M_1$ allosteric agonist, termed TBPB, was reported [201]. Interestingly, this compound was applied in *in vivo* studies, where it was shown that TBPB possessed atypical antipsychotic-like properties, in Fos-like immunoreactivity induction pattern as well as in behavioral model of psychosis, without occupying central dopamine $D_2$ receptors. Furthermore, TBPB potentiated NMDA receptor currents in hippocampal pyramidal cells through activation of $M_1$ mAChR subtype, and had effects on the processing of the amyloid precursor protein toward the non-amyloidogenic pathway and decreased $A\beta$ production *in vitro*.

Another allosteric agonist with a high selectivity at $M_1$ mAChR subtype, 77-LH-28-1, is a compound synthesized as a structural analog of AC-42 [202, 203]. It has been reported that cell firing is increased by 77-LH-28, but not by AC-42, through $M_1$ mAChR activation, in hippocampal slice experiments. Furthermore, this compound induced gamma frequency oscillation in hippocampus *in vitro*, and following subcutaneous injection *in vivo*, cell firing was facilitated in hippocampus at a dose of 3 mg/kg, in a scopolamine-sensitive manner [202]. Interestingly, 77-LH-28 and AC-42 were capable to activate $M_1$ mAChR-mediated $G\alpha_{q/11}$ and $G\alpha_s$, but not $G\alpha_{i1/2}$, whereas orthosteric agonists were activators in all of these three kinds of coupling [203]. These findings may be relevant to the possibility of development of newer drugs with functional selectivity, which act at only one of the multiple signaling pathways mediated by the same receptor subtype [204].

Bradley *et al.* [205] reported another selective $M_1$ receptor allosteric agonist AC-260584, which was orally bioavailable. This compound showed functional selectivity to $M_1$ mAChR subtype over other mAChRs in several *in vitro* biochemical assays. Furthermore, this compound activated extracellular signal-regulated kinase 1 and 2 (ERK1/2) phosphorylation in discrete mouse brain regions, mediated through $M_1$ mAChR. Interestingly, AC-260584 improved the cognitive performance in the novel object recognition test in mice, in a pirenzepine-sensitive manner.

In addition to the novel allosteric agonists of $M_1$ mAChR, many positive allosteric potentiators of $M_1$ mAChR have been recently identified by high-throughput screening approach [206-208]. Ma *et al.* [206] reported that benzyl quinolone carboxylic acid (BQCA) acted as a highly selective positive allosteric potentiator of $M_1$ mAChR *in vitro* as well as *in vivo*. When administered systematically in rodents, BQCA reversed scopolamine-induced memory deficits in contextual fear conditioning, increased blood flow to the cerebral cortex, repressed amphetamine-induce locomotion, and increased wakefulness with reduction of delta sleep. In addition, BQCA and its analogues potentiated acetylcholine-induced β-arrestin recruitment, with $EC_{50}$ values correlated with the potencies determined in calcium mobilization assay. In contrast, the allosteric agonist of $M_1$ mAChR, such as AC-42 and TBPB, failed to reverse scopolamine-induced memory deficits, and to induce β-arrestin recruitment. The pharmacological, electrophysiological, and behavioral characteristics of BQCA were also investigated in detail by Shirey *et al.* [209]. It was reported that activation of $M_1$ receptor by BQCA resulted in induction of robust inward current, increase in spontaneous EPSPs, and increased firing in medial prefrontal cortex pyramidal cells. Also, this compound restored discrimination reversal learning in a transgenic mouse model of Alzheimer's disease *in vivo*, and regulated non-amyloidogenic APP processing *in vitro*. In the report of Mario *et al.* [207], it was shown that VU 0119498, VU 0027414, VU 0090157, and VU 0029767, induced at least 5-fold leftward shifts of the acetylcholine concentration-response curve, without showing agonist activity on their own, indicative of the properties of pure positive allosteric modulators. These molecules were, however, heterogeneous in chemical structures. Furthermore, they and strikingly different in pharmacological characteristics, such as the selectivity for $M_1$ mAChR subtype, potentiating ability for the effects of acetylcholine on mutant $M_1$ receptor, the attitudes on the stimulatory effects of an allosteric agonist TBPB, and the effectiveness as a modulator depending on the $M_1$ mAChR-coupled signaling pathways. Reid *et al.* [208] identified UV 0405652 (ML 169) as a highly selective and brain penetrant positive allosteric modulator of $M_1$ mAChR, which potentiated carbachol-mediated non-amyloidogenic APPsα release. Although further multifarious investigations are necessary, these findings provide exciting hypothesis that it will be possible to develop highly selective positive allosteric modulators of $M_1$ mAChR subtype, which are clinically available as efficacious pro-cognitive drugs for the treatment of Alzheimer's disease and psychiatric disorders.

## $M_4$ mAChR

As described before, there have been several reports indicating the importance of $M_4$ mAChR, rather than $M_1$ subtype, as a molecular target implicated in the antipsychotic-like effects of xanomeline [146-148]. Therefore, selective allosteric modulators (allosteric agonists or positive allosteric potentiators) of $M_4$ mAChR, may serve as novel effective antipsychotics devoid of direct antagonistic properties at dopamine $D_2$ receptors. Indeed, several recent reports have provided the data supporting this hypothesis. A compound termed VU 10010 was reported to act as a highly selective positive allosteric modulator of $M_4$ mAChR, and whole-cell patch clamp study indicated that it increased carbachol-induced depression of transmission at excitatory synapses in hippocampal CA1 pyramidal cells [210]. Unfortunately, this compound was not suitable for *in vivo* studies, and the same research group developed the two analogs of UV 10010, named VU 0152099 and VU 0152100 [211]. Both compounds were also potent and selective allosteric modulators of $M_4$ subtype, with no agonist activities on their own. Interestingly, they reversed amphetamine-induced hyperlocomotion in rats, when systematically administered. On the other hand, Chan *et al.* [212] reported another compound LY 2033298, which was shown to be effective in conditioned avoidance responding and prepulse inhibition models in rats, when co-administered with a sub-effective dose of oxotremorine. In this report, it was also demonstrated that the potentiating effects of LY 2033298 on oxotremorine-induced inhibition of condition avoidance responding were significantly attenuated in $M_4$ mAChR knockout mice, indicative of the involvement of $M_4$ subtype. Interestingly, the modes of potentiating action of these compounds on the acetylcholine-mediated response are not completely identical to each other. Thus, the compounds such as VU 10010 potentiate the acetylcholine's effects, by shifting the concentration-response curve leftward as well as by augmenting the maximal response elicited by acetylcholine, while they are quiescent in the absence of the orthosteric agonist [210, 211]. On the other hand, LY 2033298 shows robust allosteric agonism action in many biochemical assay systems, in addition to the potentiating effects on the affinity of $M_4$ for acetylcholine, which is likely relevant also *in vivo* [213]. Then, LY 2033298 is a so-called ago-allosteric modulator of $M_4$ mAChR. The possible implications of such subtle differences in the pharmacological modes of action in their *in vivo* effects remain to be addressed further.

## mGlu$_2$ Receptor

As already described in the section regarding the orthosteric agonists of group II mGlu receptors, a large number of preclinical and clinical studies have provided the notion that stimulation of group II mGlu receptors, in particular mGlu$_2$, may represent a novel efficacious approach to the treatment of schizophrenia as well as anxiety disorders [157, 158]. However, despite the tremendous efforts for the development of group II mGlu receptors agonists, differentiation between mGlu$_2$ and mGlu$_3$ has not been successful, and it is not yet clear whether orthosteric agonists will reach the market for broad clinical use. As an alternative way, many allosteric potentiators of mGlu$_2$ receptor have been developed in recent years [214-222]. Among these compounds, LY 487379 may be somewhat unique, since it potentiates agonist-induced response in mGlu$_2$ receptors through dual actions, *i.e.*, by shifting the concentration-response curve leftward as well as by increasing the maximum response [215].

Accumulating results in behavioral studies have shown that these selective allosteric potentiators of mGlu$_2$ receptor are promising candidate drugs with antipsychotic and/or anxiolytic properties. Thus, LY 487379 inhibited PCP- and amphetamine-induced hyperlocomotor activity in mice [219]. Interestingly, this compound was also effective in reversal of amphetamine-induced disruption of prepulse inhibition of the acoustic startle reflex, which was not mimicked by the orthosteric group II mGlu receptors agonist LY 379268. In social novelty discrimination test in the rats treated with PCP during neonatal period, LY 487379 was as effective as LY 354740, an orthosteric mGlu$_{2/3}$ agonist [167]. Pro-cognitive effects of LY 487379 were also indicated in other behavioral tests [223].

Biphenyl-indanone A (BINA) is another selective positive allosteric modulator of mGlu$_2$, which was shown to be effective in PCP-induced hyperlocomotion model, but not in amphetamine-induced hyperlocomotion [222]. Interestingly, it was reported that this compound blocked PCP-induced disruption of prepulse inhibition in mice, decreased stress-induced hyperthermia, and provided anxiolytic-like effects in elevated plus maze test [222]. BINA also attenuated 5-HT-induced increase in spontaneous excitatory postsynaptic currents in medial prefrontal cortex [224]. In addition, BINA reduced (-)-DOI-induced head twitch behavior and Fos expression in medial prefrontal cortex [224].

## CONCLUDING REMARKS

The pharmacological interventions for mental illnesses have been still unsatisfactory, both in efficacy and adverse events profiles. Although newer psychotropic drugs of second generation have successfully contributed to reduction of uncomfortable, and sometimes serious, adverse effects to some extent, there have been many clinical problems left unresolved. We have many refractory patients, who show no or only limited response to the treatment with currently available psychotropic drugs. Regarding schizophrenic disorders, this is the case even for the positive symptoms, such as hallucinations and delusions, which are usually supposed to be controlled very well by blockade of central dopamine D$_2$ receptors. Most schizophrenic patients suffer from refractory negative symptoms and cognitive impairments, which usually last to the end of their lives with deteriorating processes. As a consequence, the lifelong outcome of schizophrenia has not been dramatically improved even after the introduction of atypical antipsychotics. Similarly, pharmacotherapy for the patients with mood disorders and anxiety disorders has ample room for improvement.

At present, it may not be realistically feasible to develop novel psychotropic drugs entirely irrelevant to the classical monoamine hypotheses of mental disorders. In this sense, many of the compounds described in this chapter are positioned as improvements in, or complements to, existing psychotropic drugs. Among them, several partial agonists are already clinically available, *e.g.*, aripiprazole as a dopamine $D_2$ receptor agonist and buspirone (or tandospirone) as a 5-$HT_{1A}$ receptor agonist. For clinical utility, it may be preferable that we have several agonists with different intrinsic activities in hand. The most suitable intrinsic activity may be different individually to re-stabilize the disorganized function of the targeted GPCR into the optimal range.

Despite intensive efforts in medicinal chemistry, purely selective orthosteric agonists have not been developed yet, with regard to mAChRs and mGlu receptors. Allosteric ligands are of particular significance for such GPCRs, to gain high selectivity to a certain receptor subtype. Additionally, allosteric modulators have potential advantages to fine-tune the dysregulated or unbalanced signal transduction systems toward more physiological homeostatic states, as compared with radical interventions elicited by orthosteric ligands. Although still preliminary, several allosteric potentiators of some GPCRs appear promising as effective psychotropic drugs.

## ACKNOWLEDGEMENTS

This work was financially supported by the Grant for Research Work from the Saitama Medical University, Japan. The author is grateful to Professor Ryoichi Toyoshima, the chair of Department of Psychiatry, Faculty of Medicine, Saitama Medical University, for his support and encouragement.

## CONFLICT OF INTEREST

The author confirm that this chapter has no conflict of interest.

## REFERENCES

[1]    Volz H-P, Stoll KD. Clinical trials with sigma ligands. Pharmacopsychiatry 2004; 37(Suppl 3): S214-S20.
[2]    Hayashi T, Tsai S-Y, Mori T, Fujimoto M, Su T-P. Targeting ligand-operated chaperone sigma-1 receptors in the treatment of neuropsychiatric disorders. Expert Opin Ther Targets 2011; 15(5): 557-77.

[3]     Skuza G. Pharmacology of sigma ($\sigma$) receptor ligands from a behavioral perspective. Curr Pharm Design 2012; 18(7): 863-74.

[4]     Gründer G, Müller MJ, Andrea J, Heydari N, Wetzel H, Schlösser R *et al*. Occupancy of striatal $D_2$-like dopamine receptors after treatment with the sigma ligand EMD 5744, a putative atypical antipsychotic. Psychopharmacology (Berl) 1999; 146(1): 81-6.

[5]     Zhu BT. Mechanistic explanation for the unique pharmacological properties of receptor partial agonists. Biomed Pharmacother 2005; 59(3): 76-89.

[6]     Ohlsen RI, Pilowsky LS. The place of partial agonism in psychiatry: Recent developments. J Psychopharmacol 2005; 19(4): 408-13.

[7]     Frishman WH, Saunders E. $\beta$-Adrenergic blockers. J Clin Hypertens (Greenwich) 2011; 13(9): 649-53.

[8]     Perez-Lloret S, Rascol O. Dopamine receptor agonists for the treatment of early or advanced Parkinson's disease. CNS Drugs 2010; 24(11): 941-68.

[8]     Al-Hasani R, Bruchas MR. Molecular mechanisms of opioid receptor-dependent signaling and behavior. Anesthesiology 2011; 115(6): 363-81.

[10]    O'Connor PG, Fiellin DA. Pharmacologic treatment of heroin-dependent patients. Ann Intern Med 2000; 133(1): 40-54.

[11]    Davids E, Gastpar M. Buprenorphine in the treatment of opioid dependence. Eur Neuropsychopharmacology 2004; 14(3): 209-16.

[12]    Taylor DP. Buspiron, a new approach to the treatment of anxiety. FASEB J 1988; 2(9): 2445-52.

[13]    Fulton B, Brogden RN. Buspirone. An updated review of its clinical pharmacology and therapeutic applications. CNS Drugs 1997; 7(1): 68-88.

[14]    Gilchrist A. Modulating G-protein-coupled receptors: From traditional pharmacology to allosterics. Trends Pharmacol Sci 2007; 28(8): 431-7.

[15]    Christopoulos A. Allosteric binding sites on cell-surface receptors: Novel targets for drug discovery. Nat Rev Drug Discov 2002; 1(3): 198-210.

[16]    Bridges TM, Lindsley CW. G-Protein-coupled receptors: From classical modes of modulation to allosteric mechanisms. ACS Chem Biol 2008; 3(9): 530-41.

[17]    Conn PJ, Christopoulos A, Lindsley CW. Allosteric modulators of GPCRs: A novel approach for the treatment of CNS disorders. Nat Rev Drug Discov 2009; 8(1): 41-53.

[18]    Wang L, Martin B, Brenneman R, Luttrell LM, Maudsley S. Allosteric modulators of G protein-coupled receptors: Future therapeutics for complex physiological disorders. J Pharmacol Exp Ther 2009; 331(2): 340-8.

[19]    May LT, Leach K, Sexton PM, Christopoulos A. Allosteric modulation of G protein-coupled receptors. Annu Rev Pharmacol Toxicol 2007; 47: 1-51.

[20]    Odagaki Y, Kinoshita M, Toyoshima R. Pharmacological characterization of $M_1$ muscarinic acetylcholine receptor-mediated Gq activation in rat cerebral cortical and hippocampal membranes. Naunyn Schmiedebergs Arch Pharmacol (in press).

[21]    Christopoulos A, Kenakin T. G protein-coupled receptor allosterism and complexing. Pharmacol Rev 2002; 54(2): 323-74.

[22]    Ritzén A, Mathiesen JM, Thomsen C. Molecular pharmacology and therapeutic prospects of metabotropic glutamate receptor allosteric modulators. Basic Clin Pharmacol Toxicol 2005; 97(4): 202-13.

[23]    Sur C, Mallorga PJ, Wittmann M, Jacobson MA, Pascarella D, Williams JB *et al*. *N*-Desmethylclozapine, an allosteric agonist at muscarinic 1 receptor, potentiates *N*-methyl-D-aspartate receptor activity. Proc Natl Acad Sci USA 2003; 100(23): 13674-79.

[24]  Taylor DP. Serotonin agents in anxiety. Ann NY Acad Sci 1990; 600: 545-57.

[25]  Nishitsuji K, To H, Murakami Y, Kodama K, Kobayashi D, Yamada T *et al*. Tandospirone in the treatment of generalized anxiety disorder and mixed anxiety-depression. Clin Drug Invest 2004; 24(2): 121-6.

[26]  Hamik A, Oksenberg D, Fischette C, Peroutka SJ. Analysis of tandospirone (SM-3997) interactions with neurotransmitter receptor binding sites. Biol Psychiatry 1990; 28(2): 99-109.

[27]  Tanaka H, Tatsuno T, Shimizu H, Hirose A, Kumasaka Y, Nakamura M. Effects of tandospirone on second messenger systems and neurotransmitter release in the rat brain. Gen Pharmacol 1995; 26(8): 1765-72.

[28]  Odagaki Y, Toyoshima R, Yamauchi T. Trazodone and its active metabolite *m*-chlorophenylpiperazine as partial agonists at $5\text{-HT}_{1A}$ receptors assessed by $[^{35}\text{S}]\text{GTP}\gamma\text{S}$ binding. J Psychopharmacol 2005; 19(3): 235-41.

[29]  Rickels K, Rynn M. Pharmacotherapy of generalized anxiety disorder. J Clin Psychiatry 2002; 63(Suppl 14): 9-16.

[30]  Gorman JM. Treating generalized anxiety disorder. J Clin Psychiatry 2003; 64(Suppl 2): 24-9.

[31]  Robinson D, Rickels K, Feighner J, Fabre LF, Gammans RE, Shrotriya RC *et al*. Clinical effects of the $5\text{-HT}_{1A}$ partial agonists in depression: A comparative analysis of buspirone in the treatment of depression. J Clin Psychopharmaocol 1990; 10(Suppl 3): 67S-76S).

[32]  Feiger AD, Heiser JF, Shrivastava RK, Weiss KJ, Smith WT, Sitsen JM *et al*. Gepirone extended-release: New evidence for efficacy in the treatment of major depressive disorder. J Clin Psychiatry 2003; 648(3): 243-249.

[33]  Robinson DS, Sitsen JM, Gibertini M. A review of the efficacy and tolerability of immediate-release and extended-release formulations of gepirone. Clin Ther 2003; 25(6): 1618-33.

[34]  Bielski RJ, Cunningham L, Horrigan JP, Londborg PD, Smith WT, Weiss K. Gepirone extended-release in the treatment of adult outpatients with major depressive disorder: A double-blind, randomized, placebo-controlled, parallel-group study. J Clin Psychiatry 2008; 69(4): 571-7.

[35]  Lapierre YD, Silverstone P, Reesal RT, Saxena B, Turner P, Bakish D *et al*. Canadian multicenter study of three fixed doses of controlled-release ipsapirone in outpatients with moderate to severe major depression. J Clin Psychopharmacol 1998; 18(4): 268-73.

[36]  Stahl SM, Kaiser L, Roeschen J, Keppel Hesselink JM, Orazem J. Effectiveness of ipsapirone, a 5-HT-1A partial agonist, in major depressive disorder: Support for the role of 5-HT-1A receptors in the mechanism of action of serotonergic antidepressants. Int J Neuropsychopharmacol 1998; 1(1): 11-18.

[37]  Savitz J, Lucki I, Drevets WC. $5\text{-HT}_{1A}$ receptor function in major depressive disorder. Prog Neurobiol 2009; 88(1): 17-31.

[38]  Schreiber R, De Vry J. $5\text{-HT}_{1A}$ receptor ligands in animal models of anxiety, impulsivity and depression: Multiple mechanisms of action? Prog Neuropsychopharmacol Biol Psychiatry 1993; 17(1): 87-104.

[39]  Lucki I, Singh A, Kreiss DS. Antidepressant-like behavioral effects of serotonin receptor agonists. Neurosci Biobehav Rev 1994; 18(1): 85-95.

[40]  De Vry J. $5\text{-HT}_{1A}$ receptor agonists: recent developments and controversial issues. Psychopharmacology (Berl) 1995; 121(1): 1-26.

[41]    Blier P, Ward NM. Is there a role for 5-HT$_{1A}$ agonists in the treatment of depression? Biol Psychiatry 2003; 53(3): 193-203.

[42]    Heiser JF, Wilcox CS. Serotonin 5-HT$_{1A}$ receptor agonists as antidepressants. Pharmacological rationale and evidence for efficacy. CNS Drugs 1998; 10(5): 343-53.

[43]    Koek W, Vacher B, Cosi C, Assié M-B, Patoiseau J-F, Pauwels PJ *et al.* 5-HT$_{1A}$ receptor activation and antidepressant-like effects: F 13714 has high efficacy and marked antidepressant potential. Eur J Pharmacol 2001; 420(2-3): 103-12.

[44]    Handley SL, McBlane JW. 5HT drugs in animal models of anxiety. Psychopharmacology (Berl). 1993; 112(1): 13-20.

[45]    Terawaki K, Ikarashi Y, Sekiguchi K, Nakai Y, Kase Y. Partial agonistic effect of yokukansan on human recombinant serotonin 1A receptors expressed in the membranes of Chinese hamster ovary cells. J Ethnopharmacol 2010; 127(2): 306-12.

[46]    Millan MJ. Improving the treatment of schizophrenia: focus on serotonin (5-HT)$_{1A}$ receptors. J Pharmacol Exp Ther 2000; 295(3): 853-61.

[47]    Bantick RA, Deakin JF, Grasby PM. The 5-HT$_{1A}$ receptor in schizophrenia: a promising target for novel atypical neuroleptics? J Psychopharmacol 2001; 15(1): 37-46.

[48]    Jones CA, McCreary AC. Serotonergic approaches in the development of novel antipsychotics. Neuropharmacology 2008; 55(6): 1056-65.

[49]    Newman-Tancredi A. The importance of 5-HT$_{1A}$ receptor agonism in antipsychotic drug action: Rationale and perspectives. Curr Opin Investig Drugs 2010; 11(7): 802-12.

[50]    Newman-Tancredi A, Chaput C, Verriele L, Millan MJ. Clozapine is a partial agonist at cloned, human serotonin 5-HT$_{1A}$ receptors. Neuropharmacology 1996; 35(1): 119-21.

[51]    Newman-Tancredi A, Gavaudan S, Conte C, Chaput C, Touzard M, Verrièle L *et al.* Agonist and antagonist actions of antipsychotic agents at 5-HT$_{1A}$ receptors: a [$^{35}$S]GTPγS binding study. Eur J Pharmacol 1998; 355(2-3): 245-56.

[52]    Newman-Tancredi A, Assié M-B, Leduc N, Ormière A-M, Danty N, Cosi C. Novel antipsychotics activate recombinant human and native rat serotonin 5-HT1A receptors: Affinity, efficacy and potential implications for treatment of schizophrenia. Int J Neuropsychopharmacol 2005; 8(3): 341-56.

[53]    Bruins Slot LA, De Vries L, Newman-Tancredi A, Cussac D. Differential profile of antipsychotics at serotonin 5-HT$_{1A}$ and dopamine D$_{2S}$ receptors coupled to extracellular signal-regulated kinase. Eur J Pharmacol 2006; 534(1-3): 63-70.

[54]    Odagaki Y, Toyoshima R. 5-HT$_{1A}$ receptor agonist properties of antipsychotics determined by [$^{35}$S]GTPγS binding in rat hippocampal membranes. Clin Exp Pharmacol Physiol 2007; 34(5-6): 462-6.

[55]    Meltzer HY, Sumiyoshi T. Does stimulation of 5-HT$_{1A}$ receptors improve cognition in schizophrenia? Behav Brain Res 2008; 195(1): 98-102.

[56]    Green MF, Kern RS, Braff DL, Mintz J. Neurocognitive deficits and functional outcome in schizophrenia: are we measuring the "right stuff"? Schizophr Bull 2000; 26(1): 119-36.

[57]    Reynolds GP, Kirk SL. Metabolic side effects of antipsychotic drug treatment – pharmacological mechanisms. Pharmacol Ther 2010; 125(1): 169-79.

[58]    Wood M. Role of the 5-HT$_{2C}$ receptor in atypical antipsychotics: Hero or villain? Curr Med Chem – Cent Nerv Syst Agents 2005: 5(1); 63-6.

[59]    Dunlop J, Marquis KL, Lim HK, Leung L, Kao J, Cheesman C *et al.* Pharmacological profile of the 5-HT$_{2C}$ receptor agonist WAY-163909; therapeutic potential in multiple indications. CNS Drug Rev 2006; 12(3-4): 167-77.

[60]   Marquis KL, Sabb AL, Logue SF, Brennan JA, Piesla MJ, Comery TA *et al.* WAY-163909 [(7*b*R,10*a*R)-1,2,3,4,8,9,10,10*a*-octahydro-7*bH*-cyclopenta-[*b*][1,4]diazepino[6,7,1hi]indol e]: A novel 5-hydroxytryptamine 2C receptor-selective agonist with preclinical antipsychotic-like activity. J Pharmacol Exp Ther 2007; 320(1): 486-96.

[61]   Rosenzweig-Lipson S, Sabb A, Stack G, Mitchell P, Lucki I, Malberg JE *et al.* Antidepressant-like effects of the novel, selective, 5-HT2C receptor agonist WAY-163909 in rodents. Psychopharmacology (Berl) 2007; 192(2): 159-70.

[62]   Moreau JL, Bös M, Jenck F, Martin JR, Mortas P, Wichmann J. 5HT$_{2C}$ receptor agonists exhibit antidepressant-like properties in the anhedonia model of depression in rats. Eur Neuropsychopharmacol 1996; 6(3): 169-75.

[63]   Martin JR, Bös M, Jenck F, Moreau J-L, Mutel V, Sleight AJ *et al.* 5-HT$_{2C}$ receptor agonists: Pharmacological characteristics and therapeutic potential. J Pharmacol Exp Ther 1998; 286(2): 913-24.

[64]   Cryan JF, Lucki I. Antidepressant-like behavioral effects mediated by 5-Hydroxytryptamine$_{2C}$ receptors. J Pharmacol Exp Ther 2000; 295(3): 1120-6.

[65]   Dekeyne A, Mannoury la Cour C, Gobert A, Brocco M, Lejeune F, Serres F *et al.* S32006, a novel 5-HT$_{2C}$ receptor antagonist displaying broad-based antidepressant and anxiolytic properties in rodent models. Psychopharmacology (Berl) 2008; 199(4): 549-68.

[66]   Carney RM, Shelton RC. Agomelatine for the treatment of major depressive disorder. Expert Opin Pharmacother 2011; 12(15): 2411-9.

[67]   Sharpley AL, Rawlings NB, Brain S, McTavish SF, Cowen PJ. Does agomelatine block 5-HT$_{2C}$ receptors in humans? Psychopharmacology (Berl) 2011; 213(2-3): 653-5.

[68]   Norman TR. The effect of agomelatine on 5HT$_{2C}$ receptors in humans: A clinically relevant mechanism? Psychopharmacology (Berl) 2012; 221(1):177-8.

[69]   Millan MJ, Marin P, Kamal M, Jockers R, Chanrion B, Labasque M *et al.* The melatonergic agonist and clinically active antidepressant, agomelatine, is a neutral antagonist at 5-HT$_{2C}$ receptors. Int J Neuropsychopharmacol 2011; 14(6): 768-83.

[70]   Roth BL, Hanizavareh SM, Blum AE. Serotonin receptors represent highly favorable molecular targets for cognitive enhancement in schizophrenia and other disorders. Psychopharmacology (Berl) 2004; 174(1): 17-24.

[71]   King MV, Marsden CA, Fone KC. A role for the 5-HT$_{1A}$, 5-HT$_4$ and 5-HT$_6$ receptors in learning and memory. Trends Pharmacol Sci 2008; 29(9): 482-92.

[72]   Lucas G, Rymar VV, Du J, Mnie-Filali O, Bisgaard C, Manta S *et al.* Serotonin$_4$ (5-HT$_4$) receptor agonists are putative antidepressants with a rapid onset of action. Neuron 2007; 55(5): 712-25.

[73]   Pascual-Brazo J, Castro E, Díaz Á, Valdizán EM, Pilar-Cuéllar F, Vidal R *et al.* Modulation of neuroplasticity pathways and antidepressant-like behavioural responses following the short-term (3 and 7 days) administration of the 5-HT$_4$ receptor agonist RS67333. Int J Neuropsychopharmacol 2012; 15(5): 631-43.

[74]   Licht CL, Knudsen GM, Sharp T. Effects of the 5-HT$_4$ receptor agonist RS67333 and paroxetine on hippocampal extracellular 5-HT levels. Neurosci Lett 2010; 476(2): 58-61.

[75]   Carr GV, Schechter LE, Lucki I. Antidepressant and anxiolytic effects of selective 5-HT$_6$ receptor agonists in rats. Psychopharmacology (Berl) 2011; 213(2-3): 499-507.

[76]   Svenningsson P, Tzavara ET, Qi H, Carruthers R, Witkin JM, Nomikos GG *et al.* Biochemical and behavioral evidence for antidepressant-like effects of 5-HT$_6$ receptor stimulation. J Neurosci 2007; 27(15): 4201-9.

[77]    Schechter LE, Lin Q, Smith DL, Zhang G, Shan Q, Platt B *et al*. Neuropharmacological profile of novel and selective 5-HT$_6$ receptor agonists: WAY-181187 and WAY-208466. Neuropsychopharmacology 2008; 33(6): 1323-35.

[78]    Burnham KE, Baxter MG, Bainton JR, Southam E, Dawson LA, Bannerman DM *et al*. Activation of 5-HT$_6$ receptors facilitates attentional set shifting. Psychopharmacology (Berl) 2010; 208(1): 13-21.

[79]    Kendall I, Slotten HA, Codony X, Burgueño J, Pauwels PJ, Vela JM *et al*. E-6801, a 5-HT$_6$ receptor agonist, improves recognition memory by combined modulation of cholinergic and glutamatergic neurotransmission in the rat. Psychopharmacology (Berl) 2011; 213(2-3): 413-30.

[80]    Fone KC. An update on the role of the 5-hydroxytryptamine6 receptor in cognitive function. Neuropharmacology 2008; 55(6): 1015-22.

[81]    Borsini F, Bordi F, Riccioni T. 5-HT$_6$ pharmacology inconsistencies. Pharmacol Biochem Behav. 2011; 98(2): 169-72.

[82]    Yun HM, Rhim H. The serotonin-6 receptor as a novel therapeutic target. Exp Neurobiol. 2011; 20(4): 159-68.

[83]    Roth BL, Craigo SC, Choudhary MS, Uluer A, Monsma FJ Jr, Shen Y *et al*. Binding of typical and atypical antipsychotic agents to 5-hydroxytryptamine-6 and 5-hydroxytryptamine-7 receptors. J Pharmacol Exp Ther 1994; 268(3): 1403-10.

[84]    Sebben M, Ansanay H, Bockaert J, Dumuis A. 5-HT$_6$ receptors positively coupled to adenylyl cyclase in striatal neurones in culture. Neuroreport 1994; 5(18): 2553-7.

[85]    Kohen R, Metcalf MA, Khan N, Druck T, Huebner K, Lachowicz JE *et al*. Cloning, characterization, and chromosomal localization of a human 5-HT$_6$ serotonin receptor. J Neurochem 1996; 66(1): 47-56.

[86]    Davis KL, Kahn RS, Ko G, Davidson M. Dopamine in schizophrenia: a review and reconceptualization. Am J Psychiatry 1991; 148(11): 1474-86.

[87]    Okubo Y, Suhara T, Suzuki K, Kobayashi K, Inoue O, Terasaki O *et al*. Decreased prefrontal dopamine D1 receptors in schizophrenia revealed by PET. Nature 1997; 385(6617): 634-6.

[88]    Abi-Dargham A, Moore H. Prefrontal DA transmission at D$_1$ receptors and the pathology of schizophrenia. Neuroscientist 2003; 9(5): 404-16.

[89]    Castner SA, Williams GV, Goldman-Rakic PS. Reversal of antipsychotic-induced working memory deficits by short-term dopamine D1 receptor stimulation. Science 2000; 287(5460): 2020-2.

[90]    McLean SL, Idris NF, Woolley ML, Neill JC. D$_1$-like receptor activation improves PCP-induced cognitive deficits in animal models: Implications for mechanisms of improved cognitive function in schizophrenia. Eur Neuropsychopharmacol 2009; 19(6): 440-50.

[91]    Roberts BM, Seymour PA, Schmidt CJ, Williams GV, Castner SA. Amelioration of ketamine-induced working memory deficits by dopamine D1 receptor agonists. Psychopharmacology (Berl) 2010; 210(3): 407-18.

[92]    Salmi P, Isacson R, Kull B. Dihydrexidine – the first full dopamine D$_1$ receptor agonist. CNS Drug Rev 2004; 10(3): 230-42.

[93]    Jin G-Z, Zhu Z-T, Fu Y. (-)-Stepholidine: A potential novel antipsychotic drug with dual D1 receptor agonist and D2 receptor antagonist actions. Trends Pharmacol Sci 2002; 23(1): 4-7.

[94]    Mo J, Guo Y, Yang Y-S, Shen J-S, Jin G-Z, Zhen X. Recent developments in studies of *l*-stepholidine and its analogs: chemistry, pharmacology and clinical implications. Curr Med Chem 2007; 14(28): 2996-3002.

[95]    Natesan S, Reckless GE, Barlow KBL, Odontiadis J, Nobrega JN, Baker GB *et al*. The antipsychotic potential of l-stepholidine – a naturally occurring dopamine receptor $D_1$ agonist and $D_2$ antagonist. Psychopharmacology (Berl) 2008; 199(2): 275-89.

[96]    Guo Y, Zhang H, Chen X, Cai W, Cheng J, Yang Y *et al*. Evaluation of the antipsychotic effect of bi-acetylated *l*-stepholidine (*l*-SPD-A), a novel dopamine and serotonin receptor dual ligand. Schizophr Res 2009; 115(1): 41-9.

[97]    Ahlenius S. Clozapine: dopamine $D_1$ receptor agonism in the prefrontal cortex as the code to decipher a Rosetta stone of antipsychotic drugs. Pharmacol Toxicol 1999; 84(5): 193-6.

[98]    Tamminga CA. Partial dopamine agonists in the treatment of psychosis. J Neural Transm 2002; 109(3): 411-20.

[99]    Benkert O, Müller-Siecheneder F, Wetzel H. Dopamine agonists in schizophrenia: A review. Eur Neuropsychopharmacol 1995; 5(Suppl): 43-53.

[100]   Stahl SM. Dopamine system stabilizers, aripiprazole, and the next generation of antipsychotics, part 1, "Goldilocks" actions at dopamine receptors. J Clin Psychiatry 2001; 62(11): 841-2.

[101]   Stahl SM. Dopamine system stabilizers, aripiprazole, and the next generation of antipsychotics, part 2: Illustrating their mechanism of action. J Clin Psychiatry 2001; 62(12): 923-4.

[102]   Croxtall JD. Aripiprazole. A review of its use in the management of schizophrenia in adults. CNS Drugs 2012; 26(2): 155-83.

[103]   Odagaki Y. Partial dopamine receptor agonists as newer atypical antipsychotics: Intrinsic activity appropriate for treatment of schizophrenic patients. Cent Nerv Syst Agents Med Chem 2007; 7(3): 156-65.

[104]   Cosi C, Carilla-Durand E, Assié MB, Ormiere AM, Maraval M, Leduc N *et al*. Partial agonist properties of the antipsychotics SSR181507, aripiprazole and bifeprunox at dopamine D2 receptors: G protein activation and prolactin release. Eur J Pharmacol 2006; 535(1-3): 135-44.

[105]   Heusler P, Newman-Tancredi A, Castro-Fernandez A, Cussac D. Differential agonist and inverse agonist profile of antipsychotics at $D_{2L}$ receptors coupled to GIRK potassium channels. Neuropharmacology 2007; 52(4): 1106-13.

[106]   Tadori Y, Kitagawa H, Forbes RA, McQuade RD, Stark A, Kikuchi T. Differences in agonist/antagonist properties at human dopamine $D_2$ receptors between aripiprazole, bifeprunox and SDZ 208-912. Eur J Pharmacol 2007; 574(2-3): 103-11.

[107]   Casey DE, Sands EE, Heisterberg J, Yang HM. Efficacy and safety of bifeprunox in patients with an acute exacerbation of schizophrenia: Results from a randomized, double-blind, placebo-controlled, multicenter, dose-finding study. Psychopharmacology (Berl) 2008; 200(3): 317-31.

[108]   Burstein ES, Ma J, Wong S, Gao Y, Pham E, Knapp AE *et al*. Intrinsic efficacy of antipsychotics at human $D_2$, $D_3$, and $D_4$ dopamine receptors: Identification of the clozapine metabolite *N*-desmethylclozapine as a $D_2/D_3$ partial agonist. J Pharmacol Exp Ther 2005; 315(3): 1278-87.

[109]   Bishara D, Taylor D. Upcoming agents for the treatment of schizophrenia. Mechanism of action, efficacy and tolerability. Drugs 2008; 68(16): 2269-92.

[110]   Mendoza MC, Lindenmayer JP. *N*-Desmethylclozapine: Is there evidence for its antipsychotic potential? Clin Neuropharmacol 2009; 32(3): 154-7.

[111] Parsons CG, Danysz W, Quack G. Memantine is a clinically well tolerated N-methyl-D-aspartate (NMDA) receptor antagonist – a review of preclinical data. Neuropharmacology 1999; 38(6): 735-67.

[112] Seeman P, Caruso C, Lasaga M. Memantine agonist action at dopamine $D2^{High}$ receptors. Synapse 2008; 62(2): 149-53.

[113] Witkin J, Gasior M, Acri J, Beekman M, Thurkauf A, Yuan J *et al.* Atypical antipsychotic-like effects of the dopamine $D_3$ receptor agonist, (+)-PD 128,907. Eur J Pharmacol 1998; 347(2-3): R1-3.

[114] Pugsley TA, Davis MD, Akunne HC, MacKenzie RG, Shih YH, Damsma G *et al.* Neurochemical and functional characterization of the preferentially selective dopamine D3 agonist PD 128907. J Pharmacol Exp Ther 1995; 275(3): 1355-66.

[115] Witkin JM, Levant B, Zapata A, Kaminski R, Gasior M. The dopamine $D_3/D_2$ agonist (+)-PD-128,907 [(*R*-(+)-*trans*-3,4*a*,10*b*-tetrahydro-4-propyl-2*H*,5*H*-[1]benzopyrano[4,3-*b*]-1,4-oxazin-9-ol)] protects against acute and cocaine-kindled seizures in mice: Further evidence for the involvement of $D_3$ receptors. J Pharmacol Exp Ther 2008; 326(3): 930-8.

[116] Kiss B, Horváth A, Némethy Z, Schmidt E, Laszlovszky I, Bugovics G *et al.* Cariprazine (RGH-188), a dopamine $D_3$ receptor-preferring $D_3/D_2$ dopamine receptor antagonist – partial agonist antipsychotic candidate: *In vitro* and neurochemical profile. J Pharmacol Exp Ther 2010; 333(1): 328-40.

[117] Tadori Y, Forbes RA, McQuade RD, Kikuchi T. *In vitro* pharmacology of aripiprazole, its metabolite and experimental dopamine partial agonists at human dopamine $D_2$ and $D_3$ receptors. Eur J Pharmacol 2011; 668(3): 355-65.

[118] Gyertyán I, Kiss B, Sághy K, Laszy J, Szabó G, Szabados T *et al.* Cariprazine (RGH-188), a potent $D_3/D_2$ dopamine receptor partial agonist, binds to dopamine $D_3$ receptors *in vivo* and shows antipsychotic-like and precognitive effects in rodents. Neurochem Int 2011; 59(6): 925-35.

[119] Ágai-Csongor E, Domány G, Nógrádi K, Galambos J, Vágó I, Keserű GM *et al.* Discovery of cariprazine (RGH-188): A novel antipsychotic acting on dopamine $D_3/D_2$ receptors. Bioorg Med Chem Lett 2012; 22(10): 3437-40.

[120] Gründer G. Cariprazine, an orally active $D_2/D_3$ receptor antagonist, for the potential treatment of schizophrenia, bipolar mania and depression. Curr Opin Investig Drugs 2010; 11(7): 823-32.

[121] Koch HJ, Haas S, Jürgens T. On the physiological relevance of muscarinic acetylcholine receptors in Alzheimer's disease. Curr Med Chem 2005; 12(24): 2915-21.

[122] Scarr E. Muscarinic receptors in psychiatric disorders – can we mimic 'health'? Neurosignals 2009; 17(4): 298-310.

[123] Bymaster FP, Felder C, Ahmed S, McKinzie D. Muscarinic receptors as a target for drugs treating schizophrenia. Curr Drug Targets CNS Neurol Disord 2002; 1(2): 163-81.

[124] Langmead CJ, Watson J, Reavill C. Muscarinic acetylcholine receptors as CNS drug targets. Pharmacol Ther 2008; 117(2): 232-43.

[125] Lieberman JA, Javitch JA, Moore H. Cholinergic agonists as novel treatments for schizophrenia: The promise of rational drug development for psychiatry. Am J Psychiatry 2008; 165(8): 931-6.

[126] Heinrich JN, Butera JA, Carrick T, Kramer A, Kowal D, Lock T *et al.* Pharmacological comparison of muscarinic ligands: Historical *versus* more recent muscarinic $M_1$-preferring receptor agonists. Eur J Pharmacol 2009; 605(1-3): 53-6.

[127]   Fisher A. Therapeutic strategies in Alzheimer's disease: M1 muscarinic agonists. Jpn J Pharmacol 2000; 84(2): 101-12.

[128]   McArthur RA, Gray J, Schreiber R. Cognitive effects of muscarinic $M_1$ functional agonists in non-human primates and clinical trials. Curr Opin Investig Drugs 2010; 11(7): 740-60.

[129]   Bruno G, Mohr E, Gillespie M, Fedio P, Chase TN. Muscarinic agonist therapy of Alzheimer's disease. A clinical trial of RS-86. Arch Neurol 1986; 43(7): 659-61.

[130]   Mouradian MM, Mohr E, Williams JA, Chase TN. No response to high-dose muscarinic agonist therapy in Alzheimer's disease. Neurology 1988; 38(4): 606-8.

[131]   Fisher A. Muscarinic receptor agonists in Alzheimer's disease. More than just symptomatic treatment? CNS Drugs 1999; 12(3): 197-214.

[132]   Shannon HE, Bymaster FP, Calligaro DO, Greenwood B, Mitch CH, Sawyer BD *et al.* Xanomeline: A novel muscarinic receptor agonist with functional selectivity for $M_1$ receptors. J Pharmacol Exp Ther 1994; 269(1): 271-81.

[133]   Bymaster FP, Whitesitt CA, Shannon HE, DeLapp N, Ward JS, Calligaro DO *et al.* Xanomeline: A selective muscarinic agonist for the treatment of Alzheimer's disease. Drug Dev Res 1997; 40(2): 158-70.

[134]   Bodick NC, Offen WW, Levey AI, Cutler NR, Gauthier SG, Satlin A *et al.* Effects of xanomeline, a selective muscarinic receptor agonist, on cognitive function and behavioral symptoms in Alzheimer disease. Arch Neurol 1997; 54(4): 465-73.

[135]   Bodick NC, Offen WW, Shannon HE, Satterwhite J, Lucas R, van Lier R *et al.* The selective muscarinic agonist xanomeline improves both the cognitive deficits and behavioral symptoms of Alzheimer disease. Alzheimer Dis Assoc Disord 1997; 11(Suppl 4): S16-22.

[136]   Mirza NR, Peters D, Sparks RG. Xanomeline and the antipsychotic potential of muscarinic receptor subtype selective agonists. CNS Drug Rev 2003; 9(2): 159-86.

[137]   Wood MD, Murkitt KL, Ho M, Watson JM, Brown F, Hunter AJ *et al.* Functional comparison of muscarinic partial agonists at muscarinic receptor subtypes $hM_1$, $hM_2$, $hM_3$, $hM_4$ and $hM_5$ using microphysiometry. Br J Pharmacol 1999; 126(7): 1620-4.

[138]   Jones CK, Byun N, Bubser M. Muscarinic and nicotinic acetylcholine receptor agonists and allosteric modulators for the treatment of schizophrenia. Neuropsychopharmacology 2012; 37(1): 16-42.

[139]   Shannon HE, Rasmussen K, Bymaster FP, Hart JC, Peters SC, Swedberg MDB *et al.* Xanomeline, an $M_1$/$M_4$ preferring muscarinic cholinergic receptor agonist, produces antipsychotic-like activity in rats and mice. Schizophr Res 2000; 42(3): 249-59.

[140]   Perry KW, Nisenbaum LK, George CA, Shannon HE, Felder CC, Bymaster FP. The muscarinic agonist xanomeline increases monoamine release and immediate early gene expression in the rat prefrontal cortex. Biol Psychiatry 2001; 49(8): 716-25.

[141]   Stanhope KJ, Mirza NR, Bickerdike MJ, Bright JL, Harrington NR, Hesselink MB *et al.* The muscarinic receptor agonist xanomeline has an antipsychotic-like profile in the rat. J Pharmacol Exp Ther 2001; 299(2): 782-92.

[142]   Shannon HE, Hart JC, Bymaster FP, Calligaro DO, DeLapp NW, Mitch CH *et al.* Muscarinic receptor agonists, like dopamine receptor antagonist antipsychotics, inhibit conditioned avoidance response in rats. J Pharmacol Exp Ther 1999; 290(2): 901-7.

[143]   Jones CK, Eberle EL, Shaw DB, McKinzie DL, Shannon HE. Pharmacologic interactions between the muscarinic cholinergic and dopaminergic systems in the modulation of prepulse inhibition in rats. J Pharmacol Exp Ther 2005; 312(3): 1055-63.

[144]  Andersen MB, Fink-Jensen A, Peacock L, Gerlach J, Bymaster F, Lundbaek JA *et al*. The muscarinic $M_1/M_4$ receptor agonist xanomeline exhibits antipsychotic-like activity in *Cebus apella* monkeys. Neuropsychopharmacology 2003; 28(6): 1168-75.

[145]  Shekhar A, Potter WZ, Lightfoot J, Lienemann J, Dubé S, Mallinckrodt C *et al*. Selective muscarinic receptor agonist xanomeline as a novel treatment approach for schizophrenia. Am J Psychiatry 2008; 165(8): 1033-9.

[146]  Woolley ML, Carter HJ, Gartlon JE, Watson JM, Dawson LA. Attenuation of amphetamine-induced activity by the non-selective muscarinic receptor agonist, xanomeline, is absent in muscarinic $M_4$ receptor knockout mice and attenuated in muscarinic $M_1$ receptor knockout mice. Eur J Pharmacol 2009; 603(1-3): 147-9.

[147]  Thomsen M, Wess J, Fulton BS, Fink-Jensen A, Caine SB. Modulation of prepulse inhibition through both $M_1$ and $M_4$ muscarinic receptors in mice. Psychopharmacology (Berl) 2010; 208(3): 401-16.

[148]  Dencker D, Wörtwein G, Weikop P, Jeon J, Thomsen M, Sager TN *et al*. Involvement of a subpopulation of neuronal $M_4$ muscarinic acetylcholine receptors in the antipsychotic-like effects of the $M_1/M_4$ preferring muscarinic receptor agonist xanomeline. J Neurosci 2011; 31(16): 5905-8.

[149]  Zorn SH, Jones SB, Ward KM, Liston DR. Clozapine is a potent and selective muscarinic $M_4$ receptor agonist. Eur J Pharmacol 1994; 269(3): R1-2.

[150]  Olianas MC, Maullu C, Onali P. Effects of clozapine on rat striatal muscarinic receptors coupled to inhibition of adenylyl cyclase activity and on the human cloned m4 receptor. Br J Pharmacol 1997; 122(3): 401-8.

[151]  Olianas MC, Maullu C, Onali P. Mixed agonist-antagonist properties of clozapine at different human cloned muscarinic receptor subtypes expressed in Chinese hamster ovary cells. Neuropsychopharmacology 1999; 20(3): 263-70.

[152]  Weiner DM, Meltzer HY, Veinbergs I, Donohue EM, Spalding TA, Smith TT *et al*. The role of M1 muscarinic receptor agonism of *N*-desmethylclozapine in the unique clinical effects of clozapine. Psychopharmacology (Berl) 2004; 177(1-2): 207-16.

[153]  Davies MA, Compton-Toth BA, Hufeisen SJ, Meltzer HY, Roth BL. The highly efficacious actions of *N*-desmethylclozapine at muscarinic receptors are unique and not a common property of either typical or atypical antipsychotic drugs: Is $M_1$ agonism a pre-requisite for mimicking clozapine's actions? Psychopharmacology (Berl) 2005; 178(4): 451-60.

[154]  Thomas DR, Dada A, Jones GA, Deisz RA, Gigout S, Langmead CJ *et al*. *N*-desmethylclozapine (NDMC) is an antagonist at the human native muscarinic $M_1$ receptor. Neuropharmacology 2010; 58(8): 1206-14.

[155]  Watt ML, Schober DA, Hitchcock S, Liu B, Chesterfield AK, McKinzie D *et al*. Pharmacological characterization of LY593093, an M1 muscarinic acetylcholine receptor-selective partial orthosteric agonist. J Pharmacol Exp Ther 2011; 338(2): 622-32.

[156]  Fell MJ, McKinzie DL, Monn JA, Svensson KA. Group II metabotropic glutamate receptor agonists and positive allosteric modulators as novel treatments for schizophrenia. Neuropharmacology 2012; 62(3): 1473-83.

[157]  Chavez-Noriega LE, Schaffhauser H, Campbell UC. Metabotropic glutamate receptors: Potential drug targets for the treatment of schizophrenia. Curr Drug Targets CNS Neurol Disord 2002; 1(3): 261-81.

[158]  Chaki S. Group II metabotropic glutamate receptor agonists as a potential drug for schizophrenia. Eur J Pharmacol 2010; 639(1-3): 59-66.

[159]  Moghaddam B, Adams BW. Reversal of phencyclidine effects by a group II metabotropic glutamate receptor agonist in rats. Science 1998; 281(5381): 1349-52.

[160]  Cartmell J, Monn JA, Schoepp DD. The metabotropic glutamate 2/3 receptor agonists LY354740 and LY379268 selectively attenuate phencyclidine *versus* d-amphetamine motor behaviors in rats. J Pharmacol Exp Ther 1999; 291(1): 161-70.

[161]  Cartmell J, Monn JA, Schoepp DD. The mGlu$_{2/3}$ receptor agonist LY379268 selectively blocks amphetamine ambulations and rearing. Eur J Pharmacol 2000; 400(2-3): 221-4.

[162]  Cartmell J, Monn JA, Schoepp DD. Attenuation of specific PCP-evoked behaviors by the potent mGlu2/3 receptor agonist, LY379268 and comparison with the atypical antipsychotic, clozapine. Psychopharmacology (Berl) 2000; 148(4): 423-9.

[163]  Clark M, Johnson BG, Wright RA, Monn JA, Schoepp DD. Effects of the mGlu2/3 receptor agonist LY379268 on motor activity in phencyclidine-sensitized rats. Pharmacol Biochem Behav 2002; 73(2): 339-46.

[164]  Nakazato A, Kumagai T, Sakagami K, Yoshikawa R, Suzuki Y, Chaki S *et al*. Synthesis, SARs, and pharmacological characterization of 2-amino-3 or 6-fluorobicyclo[3.1.0]hexane-2,6-dicarboxylic acid derivatives as potent, selective, and orally active group II metabotropic glutamate receptor agonists. J Med Chem 2000; 43(25): 4893-909.

[165]  Rorick-Kehn LM, Perkins EJ, Knitowski KM, Hart JC, Johnson BG, Schoepp DD *et al*. Improved bioavailability of the mGlu2/3 receptor agonist LY354740 using a prodrug strategy: *In vivo* pharmacology of LY544344. J Pharmacol Exp Ther 2006; 316(2): 905-13.

[166]  Rorick-Kehn LM, Johnson BG, Knitowski KM, Salhoff CR, Witkin JM, Perry KW *et al. In vivo* pharmacological characterization of the structurally novel, potent, selective mGlu2/3 receptor agonist LY404039 in animal models of psychiatric disorders. Psychopharmacology (Berl) 2007; 193(1): 121-36.

[167]  Harich S, Gross G, Bespalov A. Stimulation of the metabotropic glutamate 2/3 receptor attenuates social novelty discrimination deficits induced by neonatal phencyclidine treatment. Psychopharmacology (Berl) 2007; 192(4): 511-9.

[168]  Amitai N, Markou A. Effects of metabotropic glutamate receptor 2/3 agonism and antagonism on schizophrenia-like cognitive deficits induced by phencyclidine in rats. Eur J Pharmacol 2010; 639(1-3): 67-80.

[169]  Spooren WP, Gasparini F, van der Putten H, Koller M, Nakanishi S, Kuhn R. Lack of effect of LY314582 (a group 2 metabotropic glutamate receptor agonist) on phencyclidine-induced locomotor activity in metabotropic glutamate receptor 2 knockout mice. Eur J Pharmacol 2000; 397(1): R1-2.

[170]  Fell MJ, Svensson KA, Johnson BG, Schoepp DD. Evidence for the role of metabotropic glutamate (mGlu)2 not mGlu3 receptors in the preclinical antipsychotic pharmacology of the mGlu2/3                                      receptor                                      agonist (-)-(1R,4S,5S,6S)-4-amino-2-sulfonylbicyclo[3.1.0]hexane-4,6-dicarboxylic                        acid (LY404039). J Pharmacol Exp Ther 2008; 326(1): 209-17.

[171]  Patil ST, Zhang L, Martenyi F, Lowe SL, Jackson KA, Andreev BV *et al*. Activation of mGlu2/3 receptors as a new approach to treat schizophrenia: a randomized Phase 2 clinical trial. Nat Med 2007; 13(9): 1102-7.

[172]  Kinon BJ, Zhang L, Millen BA, Osuntokun OO, Williams JE, Kollack-Walker S *et al*. A multicenter, inpatient, phase 2, double-blind, placebo-controlled dose-ranging study of

LY2140023 monohydrate in patients with *DSM-IV* schizophrenia. J Clin Psychopharmacol 2011; 31(3): 349-55.

[173]  Seeman P. Glutamate and dopamine components in schizophrenia. J Psychiatry Neurosci 2009; 34(2): 143-9.

[174]  Seeman P, Guan H-C. Glutamate agonist LY404,039 for treating schizophrenia has affinity for the dopamine $D2^{High}$ receptor. Synapse 2009; 63(10): 935-9.

[175]  Seeman P, Caruso C, Lasaga M. Dopamine partial agonist actions of the glutamate receptor agonists LY 354,740 and LY 379,268. Synapse 2008; 62(2): 154-8.

[176]  Fell MJ, Perry KW, Falcone JF, Johnson BG, Barth VN, Rash KS *et al. In vitro* and *in vivo* evidence for a lack of interaction with dopamine $D_2$ receptors by the metabotropic glutamate 2/3 receptor agonists 1*S*,2*S*,5*R*,6*S*-2-aminobicyclo[3.1.0]hexane-2,6-bicaroxylate monohydrate (LY354740) and (-)-2-oxa-4-aminobicyclo[3.1.0] hexane-4,6-dicarboxylic acid (LY379268). J Pharmacol Exp Ther 2009; 331(3): 1126-36.

[177]  Zysk JR, Widzowski D, Sygowski LA, Knappenberger KS, Spear N, Elmore CS *et al.* Absence of direct effects on the dopamine D2 receptor by mGluR2/3-selective receptor agonists LY 354,740 and LY 379,268. Synapse 2011; 65(1): 64-8.

[178]  Krystal JH, Abi-Saab W, Perry E, D'Souza DC, Liu N, Gueorguieva R *et al.* Preliminary evidence of attenuation of the disruptive effects of the NMDA glutamate receptor antagonist, ketamine, on working memory by pretreatment with the group II metabotropic glutamate receptor agonist, LY354740, in healthy human subjects. Psychopharmacology (Berl) 2005; 179(1): 303-9.

[179]  Kapur S, Seeman P. NMDA receptor antagonists ketamine and PCP have direct effects on the dopamine $D_2$ and serotonin 5-$HT_2$ receptors – implications for models of schizophrenia. Mol Psychiatry 2002 ;7(8): 837-44.

[180]  Jordan S, Chen R, Fernalld R, Johnson J, Regardie K, Kambayashi J *et al. In vitro* biochemical evidence that the psychotomimetics phencyclidine, ketamine and dizocilpine (MK-801) are inactive at cloned human and rat dopamine $D_2$ receptors. Eur J Pharmacol 2006; 540(1-3): 53-6.

[181]  Palucha A, Pilc A. Metabotropic glutamate receptor ligands as possible anxiolytic and antidepressant drugs. Pharmacol Ther 2007; 115(1): 116-47.

[182]  Schoepp DD, Wright RA, Levine LR, Gaydos B, Potter WZ. LY354740, an mGlu2/3 receptor agonist as a novel approach to treat anxiety/stress. Stress 2003; 6(3): 189-97.

[183]  Grillon C, Cordova J, Levine LR, Morgan CA 3rd. Anxiolytic effects of a novel group II metabotropic glutamate receptor agonist (LY354740) in the fear-potentiated startle paradigm in humans. Psychopharmacology (Berl) 2003; 168(4): 446-54.

[184]  Bergink V, Westenberg HGM. Metabotropic glutamate II receptor agonists in panic disorder: a double blind clinical trial with LY354740. Int Clin Psychopharmacol 2005; 20(6): 291-3.

[185]  Kellner M, Muhtz C, Stark K, Yassouridis A, Arlt J, Wiedemann K. Effects of a metabotropic glutamate$_{2/3}$ receptor agonist (LY544344/LY354740) on panic anxiety induced by cholecystokinin tetrapeptide in healthy humans: preliminary results. Psychopharmacology (Berl) 2005; 179(1): 310-5.

[186]  Dunayevich E, Erickson J, Levine L, Landbloom R, Schoepp DD, Tollefson GD. Efficacy and tolerability of an mGlu2/3 agonist in the treatment of generalized anxiety disorder. Neuropsychopharmacology 2008; 33(7): 1603-10.

[187]  Danysz W. LY-544344. Eli Lilly. IDrugs 2005; 8(9): 755-62.

[188]  Bernaerts P, Tirelli E. Facilitatory effect of the dopamine D4 receptor agonist PD168,077 on memory consolidation of an inhibitory avoidance learned response in C57BL/6J mice. Behav Brain Res 2003; 142(1-2): 41-52.

[189]  Browman KE, Curzon P, Pan JB, Molesky AL, Komater VA, Decker MW *et al.*, A-412997, a selective dopamine D$_4$ agonist, improves cognitive performance in rats. Pharmacol Biochem Behav. 2005; 82(1): 148-55.

[190]  Newman-Tancredi A, Assié MB, Martel J-C, Cosi C, Slot LB, Palmier C *et al.* F15063, a potential antipsychotic with D$_2$/D$_3$ antagonist, 5-HT $_{1A}$ agonist and D$_4$ partial agonist properties: (I) *In vitro* receptor affinity and efficacy profile. Br J Pharmacol 2007; 151(2): 237-52.

[191]  Depoortère R, Auclair AL, Bardin L, Bruins Slot L, Kleven MS, Colpaert F *et al.* A. F15063, a compound with D$_2$/D$_3$ antagonist, 5-HT $_{1A}$ agonist and D$_4$ partial agonist properties: (III) Activity in models of cognition and negative symptoms. Br J Pharmacol 2007; 151(2): 266-77.

[192]  Wierońska JM, Stachowicz K, Pałucha-Poniewiera A, Acher F, Brański P, Pilc A. Metabotropic glutamate receptor 4 novel agonist LSP1-2111 with anxiolytic, but not antidepressant-like activity, mediated by serotonergic and GABAergic systems. Neuropharmacology 2010; 59(7-8): 627-34.

[193]  Wierońska JM, Stachowicz K, Acher F, Lech T, Pilc A. Opposing efficacy of group III mGlu receptor activators, LSP1-2111 and AMN082, in animal models of positive symptoms of schizophrenia. Psychopharmacology (Berl) 2012; 220(3): 481-94.

[194]  Birdsall NJM, Farries T, Gharagozloo P, Kobayashi S, Kuonen D, Lazareno S *et al.* Selective allosteric enhancement of the binding and actions of acetylcholine at muscarinic receptor subtypes. Life Sci 1997; 60(13-14): 1047-52.

[195]  Jakubík J, Bacáková L, El-Fakahany EE, Tucek S. Positive cooperativity of acetylcholine and other agonists with allosteric ligands on muscarinic acetylcholine receptors. Mol Pharmacol 1997; 52(1): 172-9.

[196]  Lazareno S, Gharagozloo P, Kuonen D, Popham A, Birdsall NJ. Subtype-selective positive cooperative interactions between brucine analogues and acetylcholine at muscarinic receptors: radioligand binding studies. Mol Pharmacol 1998; 53(3): 573-89.

[197]  Birdsall NJM, Farries T, Gharagozloo P, Kobayashi S, Lazareno S, Sugimoto M. Subtype-selective positive cooperative interactions between brucine analogs and acetylcholine at muscarinic receptors: functional studies. Mol Pharmacol 1999; 55(4): 778-86.

[198]  Spalding TA, Trotter C, Skjaerbaek N, Messier TL, Currier EA, Burstein ES *et al.* Discovery of an ectopic activation site on the M$_1$ muscarinic receptor. Mol Pharmacol 2002; 61(6): 1297-302.

[199]  Langmead CJ, Fry VAH, Forbes IT, Branch CL, Christopoulos A, Wood MD *et al.* Probing the molecular mechanism of interaction between 4-*n*-butyl-1-[4-(2-methylphenyl)-4-oxo-1-butyl]-piperidine (AC-42) and the muscarinic M$_1$ receptor: Direct pharmacological evidence that AC-42 is an allosteric agonist. Mol Pharmacol 2006; 69(1): 236-46.

[200]  Spalding TA, Ma J-N, Ott TR, Friberg M, Bajpai A, Bradley SR *et al.* Structural requirements of transmembrane domain 3 for activation by the M$_1$ muscarinic receptor agonists AC-42, AC-260584, clozapine, and *N*-desmethylclozapine: Evidence for three distinct modes of receptor activation. Mol Pharmacol 2006; 70(6): 1974-83.

[201]   Jones CK, Brady AE, Davis AA, Xiang Z, Bubser M, Tantawy MN *et al.* Novel selective allosteric activator of the $M_1$ muscarinic acetylcholine receptor regulates amyloid processing and produces antipsychotic-like activity in rats. J Neurosci 2008; 28(41): 10422-33.

[202]   Langmead CJ, Austin NE, Branch CL, Brown JT, Buchanan KA, Davies CH *et al.* Characterization of a CNS penetrant, selective $M_1$ muscarinic receptor agonist, 77-LH-28-1. Br J Pharmacol 2008; 154(5): 1104-15.

[203]   Thomas RL, Mistry R, Langmead CJ, Wood MD, Challiss RAJ. G protein coupling and signaling pathway activation by $M_1$ muscarinic acetylcholine receptor orthosteric and allosteric agonists. J Pharmacol Exp Ther 2008; 327(2): 365-74.

[204]   Urban JD, Clarke WP, von Zastrow M, Nichols DE, Kobilka B, Weinstein H *et al.* Functional selectivity and classical concepts of quantitative pharmacology. J Pharmacol Exp Ther 2007; 320(1): 1-13.

[205]   Bradley SR, Lameh J, Ohrmund L, Son T, Bajpai A, Nguyen D *et al.* AC-260584, an orally bioavailable $M_1$ muscarinic receptor allosteric agonist, improves cognitive performance in an animal model. Neuropharmacology 2010; 58(2): 365-73.

[206]   Ma L, Seager MA, Wittmann M, Jacobson M, Bickel D, Burno M *et al.* Selective activation of the $M_1$ muscarinic acetylcholine receptor achieved by allosteric potentiation. Proc Natl Acad Sci U S A 2009; 106(37): 15950-5.

[207]   Marlo JE, Niswender CM, Days EL, Bridges TM, Xiang Y, Rodriguez AL *et al.* Discovery and characterization of novel allosteric potentiators of $M_1$ muscarinic receptors reveals multiple modes of activity. Mol Pharmacol 2009; 75(3): 577-88.

[208]   Reid PR, Bridges TM, Sheffler DJ, Cho HP, Lewis LM, Days E *et al.* Discovery and optimization of a novel, selective and brain penetrant $M_1$ positive allosteric modulator (PAM): The development of ML169, an MLPCN probe. Bioorg Med Chem Lett 2011; 21(9): 2697-701.

[209]   Shirey JK, Brady AE, Jones PJ, Davis AA, Bridges TM, Kennedy JP *et al.* A selective allosteric potentiator of the $M_1$ muscarinic acetylcholine receptor increases activity of medial prefrontal cortical neurons and restores impairments in reversal learning. J Neurosci 2009; 29(45): 14271-86.

[210]   Shirey JK, Xiang Z, Orton D, Brady AE, Johnson KA, Williams R *et al.* An allosteric potentiator of $M_4$ mAChR modulates hippocampal synaptic transmission. Nat Chem Biol 2008; 4(1): 42-50.

[211]   Brady AE, Jones CK, Bridges TM, Kennedy JP, Thompson AD, Heiman JU *et al.* Centrally active allosteric potentiators of the $M_4$ muscarinic acetylcholine receptor reverse amphetamine-induced hyperlocomotor activity in rats. J Pharmacol Exp Ther 2008; 327(3): 941-53.

[212]   Chan WY, McKinzie DL, Bose S, Mitchell SN, Witkin JM, Thompson RC *et al.* Allosteric modulation of the muscarinic $M_4$ receptor as an approach to treating schizophrenia. Proc Natl Acad Sci U S A 2008; 105(31): 10978-83.

[213]   Leach K, Loiacono RE, Felder CC, McKinzie DL, Mogg A, Shaw DB *et al.* Molecular mechanisms of action and *in vivo* validation of an $M_4$ muscarinic acetylcholine receptor allosteric modulator with potential antipsychotic properties. Neuropsychopharmacology 2010; 35(4): 855-69.

[214]   Lorrain DS, Schaffhauser H, Campbell UC, Baccei CS, Correa LD, Rowe B *et al.* Group II mGlu receptor activation suppresses norepinephrine release in the ventral hippocampus and

locomotor responses to acute ketamine challenge. Neuropsychopharmacology 2003; 28(9): 1622-32.

[215] Schaffhauser H, Rowe BA, Morales S, Chavez-Noriega LE, Yin R, Jachec C *et al.* Pharmacological characterization and identification of amino acids involved in the positive modulation of metabotropic glutamate receptor subtype 2. Mol  Pharmacol 2003; 64(4): 798-810.

[216] Johnson MP, Baez M, Jagdmann GE Jr, Britton TC, Large TH, Callagaro DO *et al.* Discovery of allosteric potentiators for the metabotropic glutamate 2 receptor: Synthesis and subtype                             selectivity                             of *N*-(4-(2-methoxyphenoxy)phenyl)-*N*-(2,2,2-trifluoroethylsulfonyl)pyrid-3-ylmethylamine. J Med Chem 2003; 46(15): 3189-92.

[217] Pinkerton AB, Vernier J-M, Schaffhauser H, Rowe BA, Campbell UC, Rodriguez DE *et al.* Phenyl-tetrazolyl acetophenones: Discovery of positive allosteric potentiatiors for the metabotropic glutamate 2 receptor. J Med Chem 2004; 47(18): 4595-9.

[218] Cube RV, Vernier J-M, Hutchinson JH, Gardner MF, James JK, Rowe BA *et al.* 3-(2-Ethoxy-4-{4-[3-hydroxy-2-methyl-4-(3-methylbutanoyl)phenoxy]butoxy}phenyl)prop anoic acid: A brain penetrant allosteric potentiator at the metabotropic glutamate receptor 2 (mGluR2). Bioorg Med Chem Lett 2005; 15(9): 2389-93.

[219] Galici R, Echemendia NG, Rodriguez AL, Conn PJ. A selective allosteric potentiator of metabotropic glutamate (mGlu) 2 receptors has effects similar to an orthosteric mGlu2/3 receptor agonist in mouse models predictive of antipsychotic activity. J Pharmacol Exp Ther 2005; 315(3): 1181-7.

[220] Govek SP, Bonnefous C, Hutchinson JH, Kamenecka T, McQuiston J, Pracitto R *et al.* Benzazoles as allosteric potentiators of metabotropic glutamate receptor 2 (mGluR2): Efficacy in an animal model for schizophrenia. Bioorg Med Chem Lett 2005; 15(18): 4068-72.

[221] Pinkerton AB, Cube RV, Hutchinson JH, James JK, Gardner MF, Rowe BA *et al.* Allosteric potentiators of the metabotropic glutamate receptor 2 (mGlu2). Part 3: Identification and biological activity of indanone containing mGlu2 receptor potentiators. Bioorg Med Chem Lett 2005; 15(6): 1565-71.

[222] Galici R, Jones CK, Hemstapat K, Nong Y, Echemendia NG, Williams LC *et al.* Biphenyl-indanone A, a positive allosteric modulator of the metabotropic glutamate receptor subtype 2, has antipsychotic- and anxiolytic-like effects in mice. J Pharmacol Exp Ther 2006; 318(1): 173-85.

[223] Nikiforuk A, Popik P, Drescher KU, van Gaalen M, Relo A-L, Mezler M *et al.* Effects of a positive allosteric modulator of group II metabotropic glutamate receptors, LY487379, on cognitive flexibility and impulsive-like responding in rats. J Pharmacol Exp Ther 2010; 335(3): 665-73.

[224] Benneyworth MA, Xiang Z, Smith RL, Garcia EE, Conn PJ, Sanders-Bush E. A selective positive allosteric modulator of metabotropic glutamate receptor subtype 2 blocks a hallucinogenic drug model of psychosis. Mol Pharmacol 2007; 72(2): 477-84.

# CHAPTER 2

# Central Glia as Prospective Targets in Chronic Pain

## Romain-Daniel Gosselin[*]

*Pain Centre, Department of Anesthesiology, University Hospital Centre (CHUV) and Department of Fundamental Neuroscience, University of Lausanne, Lausanne, Switzerland*

**Abstract:** The pharmacology against chronic, especially neuropathic, pain is largely unsatisfactory despite its high prevalence in the population, extremely debilitating nature and associated socioeconomic burden. Molecules in the current medicinal arsenal are mostly designed to symptomatologically target peripheral actors of tissue injury or components of neuronal hyperexcitability. These medicines particularly include opiates, sodium channel blockers, modulators of excitatory or inhibitory neurotransmission and anti-inflammatory drugs. However chronic pain is not merely a peripheral or neuronal condition. Indeed, a revolutionary vision has emerged over the past decade positing that amid the profound central plasticity associated with chronic pain, molecular changes occurring in glia are pivotal. A surge of interest has therefore spread in the field ultimately leading to a flourishing wealth of data with respect to the involvement of astrocytes and microglia in pathological pain in experimental models. Nevertheless, the vital roles exerted by glial cells argue in favour of the development of well-designed regulators of glial physiology rather than full inhibitors of glial physiology. In addition, despite the extent of data released regarding glial modifications in animal models, only scarce evidence point to a similar plasticity in human. The current chapter aims to expose the promising recent developments in the quest of modulators of glial activity, particularly emphasizing the hurdles embodied by their pharmacodynamics and pharmacokinetics specificities. The credibility and bench-to-hospital potential of the few disclosed emerging glia-active compounds that are currently in the pipeline are critically discussed. A foremost attention is given to the specific case of neuropathic pain owing to the important literature available in the field, but the discussion spreads to wider perspective that includes other forms of chronic pain.

**Keywords:** Chronic pain, neuropathic pain, glia, astrocytes, microglia, pharmacology, MAP Kinases, antibiotics, transporters, animal models, clinical trials.

---

*Address correspondence to Romain-Daniel Gosselin: Pain Centre, Department of Anesthesiology, University Hospital Centre (CHUV) and Department of Fundamental Neuroscience, University of Lausanne, Lausanne, Switzerland; Tel: +41 21 692 5254; Fax: +41 21 692 5275; E-mail: Romain-Daniel.Gosselin@chuv.ch

Atta-ur-Rahman (Ed)
All rights reserved-© 2013 Bentham Science Publishers

# INTRODUCTION: CHALLENGES OF CHRONIC PAIN TREATMENTS

## General Considerations

The need for pain alleviation has been long-standing and has probably surfaced alongside medicine itself in ancient times. Early descriptions of pain-killing "pharmacological" therapies under the form of herbal preparations date back to Hippocrates (ac. 460 BC - ac. 377 BC) with his seminal depiction of analgesic properties of opium poppy and willow tree (leaves and barks) [1, 2]. Since these antique descriptions many efforts aiming to purify and characterize various active compounds, especially since the late 19th century, have provided undisputable progresses in pain pharmacology and management while drastically reducing their undesirable side effects and toxicity. Increasingly refined opiates have replaced crude extracts from opium poppy, acetyl salicylic acid (aspirin) has been identified as the active compound from willow trees and the 20th century has been marked by the synthesis and widespread use of paracetamol (acetaminophen), non-steroidal anti-inflammatory drugs (NSAIDs) and lately modulators of neuronal activity, conduction and excitability. Nevertheless, when thoroughly analyzing this modern medicinal arsenal, it is conspicuous that despite the apparent diversity of molecules and formulations only very few modifications have been made over the last decades in terms of actual pharmaceutical classes of products. This is especially obvious and of particular clinical importance in the case of chronic/pathological pain where pain persists for months or years (or even become permanent), frequently outliving the resolution of the triggering insult or even emerging in the absence of any detectable lesion. In this case, pain is not a symptom anymore but rather becomes pathology *per se*, roughly characterized by three possible major sensory symptoms: allodynia (pain in response to normally non painful stimulus); hyperalgesia (increased pain sensitivity); and spontaneous pain.

As an umbrella idiom, *chronic pain* encompasses a plethora of poorly related clinical entities of diverse etiologies and affecting different anatomical sites. The corresponding medicinal compendium comprises wealth of different formulations and posologies with variable efficacies. A convenient, but likely highly simplistic, classification distinguishes nociceptive/inflammatory (considered as a consequence of an excess of nociceptive stimulation and inflammatory sensitization); neuropathic (from a lesion to the somatosensory system) and functional (ill-identified origin) chronic pain. However, conceptual and clinical boundaries are often vague in this classification for instance due to a mixed origin of pain (*e.g.,* inflammation or tumor of the nervous system). Furthermore, the clinical pictures are only partly exclusive despite significant extent of specificity. Even though in clinical practice, the identification of one type of pain is often guided by the responsiveness to presumably specific medications, one must keep in mind that different types of divergent pains may respond to similar molecules. In the following, I will simplistically often use neuropathic pain as an illustration, but for the above-mentioned reasons we may refer to other forms of pathological pain.

## The Unmet Pharmacological Need of Neuropathic Pain

Even though neuropathic pain represents a sub-division of chronic pain, it gathers many separate syndromes with different etiologies such as traumatic nerve injury, diabetic neuropathy, post-herpetic neuralgia, post-stroke pain, inflammation in the central nervous system or chemotherapy-induced neuropathy. This lack of homogeneity is a strong hurdle that hampers both the clear delineation of medicinal algorithms to be used in clinical practice, but also the conduct of reliable clinical trials for innovative strategies on standardized populations of patients. Guidelines for neuropathic pain management may deeply vary depending on the subtype of pain, but it usually includes: (i) tricyclic (TCA) and serotonin/norepinephrine (SNRI) antidepressants as first and second lines of treatment, especially for pain from diabetic, post-herpetic and stroke origins. These involve amitriptyline, desipramine, imipramine and duloxetine; (ii) channel ($Na^+$, $Ca^{2+}$) blockers/anticonvulsants such as carbamazepine, oxcabazepine (especially for trigeminal neuralgia), gabapentin, pregabalin, lamotrigene or phenytoine; (iii) and, to a lesser extent, antagonists of glutamate signaling (*e.g.,* dextromethorphan, memantine), opiates and cannabinoids. All these medications are accompanied by many incapacitating and worrisome side effects (such as sedation, dizziness, edema and ataxia) and are often used in combination therapy with each other and/or together with topical agents (such as lidocaine). Noteworthy, despite their wide utilization in neuropathic patients, which is likely due to over-the-counter obtainability and relatively limited adverse effects,

NSAIDs and paracetamol (acetaminophen) have not clearly proven efficacy in the case of neuropathic pain [3].

It is remarkable that not only is this pharmacopeia rather modest, but also it is far from fully effective at alleviating symptoms a satisfactory manner. The limited diversity of active compounds that target restricted sets of molecular elements despite the highly heterogeneous nature of neuropathic pain is probably an important explanation for the overall poor efficacy of symptom management. Indeed, electrical conduction (ion channels) and neurotransmission (monoamine reuptake, glutamate antagonizing) are practically the only sites of actions of the above-listed molecules. Naturally targeting ion channels and neurotransmitter receptors must remain an active field of research due to the critical importance of these proteins in nociception and to their important changes undergone in neuropathic pain. Another rationale behind the quest for innovative unifying pharmacological strategies should also be the identification of other cellular adaptations driving the hypersensitivity in the widest number of pains, which could be converted into credible all-embracing targets.

## THE GLIAL THEORY OF CHRONIC PAIN

As conspicuously pointed above, most, if not all, pharmacological tools employed to reduce the symptomatology of neuropathic pain are designed to modulate neuronal proteins. However, over the past decade scores of converging data have been amassed in experimental models showing that neuronal changes constitute the tip of the iceberg in the overall pain-related central nervous system (CNS) plasticity. Indeed, profound modifications affecting non-neuronal cells have been almost invariably described in experimental models including neuropathic (but also for example inflammatory, cancer and functional) pain, with the principal CNS cell types affected being microglia and astrocytes [4]. Animals experiencing neuropathic pain undergo glial phenotypical changes, particularly in the spinal cord (the first synaptic relay of the sensory pathway). Hence, the glial reaction emerged as a cornerstone mechanism accounting for the enduring hypersensitivity.

The perspective of acting on glia to reduce pain is not dissociable from the comprehensive understanding of both how these cells influence nociception and the molecular modifications they undergo during chronic pain. Astrocytes represent the most frequent cell type in the mammalian CNS and the pivotal situation of their processes that encase and modulate synapses and blood vessels make these cells critical in CNS homeostasis. Space limitation obstructs the full inventory of astrocytic functions. However, among the continuously growing number of functions fulfilled by astrocytes, the reuptake of extracellular neurotransmitters is regarded as particularly important in pain signaling. Through this activity, astrocytes maintain the extracellular concentration of fast-acting neurotransmitters (glutamate and GABA) in low range, which warrants a proper signal over background neurotransmission and avoids jeopardizing neuronal integrity. Microglia are CNS resident macrophage-related cells that exert a continual immune scrutiny of the neural parenchyma [5]. Classically considered as devoid of active function in the healthy CNS, microglia are now increasingly deemed dynamic players in CNS homeostasis where they constantly adjust their physiology to neuronal activity and participate in synaptic plasticity.

Following spinal insults, and even more strikingly after peripheral damages leading to chronic pain, astrocytes and microglia in the spinal cord undergo morphological and functional adaptations typically defined as *glial activation*. Microglial proliferation and shrinkage, astrocytic swelling, parenchymal release of pro-inflammatory mediators and altered astrocytic capacity to transport neurotransmitters are among the varieties of changes reported in rodent models of chronic pain. The comprehensive inventory of glial modifications reported in the varieties of chronic pain models was listed in good review articles [6]. Although significant discrepancies in glial responses have been exposed in the available literature, several key glial processes were recurrently put forward. First, upon abnormal neuronal firing generated as a consequence of peripheral lesions, astrocytes and microglia release myriads of pro-inflammatory mediators such as cytokines (*e.g.,* interleukin 1β; tumor necrosis factor α), chemokines (*e.g.,* C-C motif ligand 2/CCL2; C-X3-C motif ligand 1/CX3CL1), eicosanoids, proteases (*e.g.,* cathepsin S; matrix metalloprotease-2/MMP2) or nitric oxide (NO). In turn, these soluble factors act on neurons at presynaptic (increase in presynaptic discharges and neurotransmitter exocytosis) and post synaptic (increased glutamate-driven neurotransmission, weakened GABA inhibition) levels thus contributing to the reinforcement nociceptive transmission. In addition, pro-inflammatory factors exert autocrine influences on glial cells, likely generating a self-sustained loop of glial reaction. The release of active compounds by glia extends to so-called gliotransmitters, which modulate neuronal activity and synaptic strength for example glutamate, purines (such as adenosine tri-phosphate/ATP) or D-serine, a vital coactivator of glutamatergic neurotransmission.

## THE EFFICIENT BUT UNREALISTIC FULL GLIAL INHIBITION

The demonstration of a glial reaction in the spinal cord in response to peripheral lesions has generated great hopes that fully inhibiting glial function might reduce chronic pain symptoms. Historically, the Krebs cycle inhibitor fluorocitrate (or its precursor fluoroacetate) was the first to be successfully and recurrently used to shut down astrocyte metabolism and reaction in rodent models of chronic, especially neuropathic, pain [7]. However, the perspective of entirely shutting down astrocyte metabolism is simply not tenable. Indeed, if the full reduction of glial functions has proven its relevance at demonstrating glial importance in experimental pain, the vital multifaceted role played by astrocytes in neurophysiology strongly discredits the clinical use of fluorocitrate, as assessed by the toxicity displayed by this compound in human poisoning [8]. Albeit less dramatic, the situation is similar with respect to microglial inhibition. The tetracycline antibiotic minocycline has anti-inflammatory properties and has proven to be a potent inhibitor of microglial reaction and has been fruitfully utilized to prevent the behavioral manifestation of chronic pain in rodent [9]. However, if one can intuitively imagine the reduction of microglial function to be less detrimental than the inhibition of astrocytes due to the more limited microglial role in CNS homeostasis, long-term effects on CNS immunity might arise. Furthermore, the exact influence of microglia on neurophysiology is poorly described and might be more important than anticipated.

The simplistic *on-off* description of glial activation, which classically posits that astrocytes and microglia are constitutively *resting* supportive cells with the potential of becoming the driving force of pathologies upon activation, is not acceptable anymore. In the case of astrocytes and their all-embracing housekeeping regulatory actions, some functions (*e.g.,* cytokine release, increased vascular permeability) could indeed be foes to combat in chronic pain whereas others (*e.g.,* glutamate reuptake) might be beneficial. It is moreover plausible that astrocytes and microglia are composed of mixed populations of cells with divergent or opposite influence on neurotransmission. In this case, a full inhibition of glial physiology might generate an unwelcome composite outcome with a hypothetical reduced function of both pro-nociceptive and anti-nociceptive glia. More realistic strategies thus require refined approaches of inhibiting or activating selective relevant molecular glial targets specifically involved in neuropathic pain.

## INHIBITORS OF SPECIFIC GLIAL SIGNALING PATHWAYS

Recent advances have been made regarding the documentation of glia-specific pathways in chronic pain, the most promising example of which has been the description of mitogen-activated protein kinases (MAPKs). MAPK are a family of intracellular signaling proteins involved in many types of information transfer from extracellular signals to a cell response. MAPK are sub-divided into three classes named extracellular signal-regulated kinases (ERK), p38 and c-jun N-terminal kinases (JNK) whose respective spinal activations in chronic pain occur in a cell- and lesion-specific manner [10]. The full description of MAPK activation in different models is beyond the scope of the present overview, but the classical examples are the p38 MAPK activation in microglia and JNK cascade in astrocytes, which have been identified as potentially specific in neuropathic pain. Importantly, intrathecal blockades of p38 or JNK are possible and have been successfully employed to prevent and reverse behavioral symptoms in experimental rodent models of neuropathic pain and beyond. In particular, non-peptidergic inhibitors of p38 MAPK were successful at reducing inflammatory and neuropathic pain in rodents [11-14].

The efficacy of p38 MAPK inhibition in neuropathic pain has been evaluated in a double blind, placebo controlled cross over clinical trial (identification number NCT00390845) and the results were encouraging [15]. Dilmapimod (SB-681323) given twice daily orally for two weeks was effective at significantly attenuating pain symptoms in neuropathic patients. Other clinical trials targeting p38 have been performed in neuropathic pain (NCT00969059, NCT01110057), using losmapimod (GW856553) but the results were not significant between treated and placebo groups (Table **1**).

**Table 1:** Summary of ongoing or completed clinical trials involving glia-active compounds.

| NAME(S) | TARGET | STAGE, SPONSOR (CLINICAL TRIAL N° IF AVAILABLE) | RESULTS ON PAIN ([REF.] IF PUBLISHED) |
|---|---|---|---|
| **DILMAPIMOD (SB-681323)** | p38 MAPK | Phase II GlaxoSmithKline (GSK protocol Id: MKN106762) (NCT00390845) | Reduced average daily pain *vs*. placebo. Well tolerated [14] |
| **LOSMAPIMOD (GW-856553)** | p38 MAPK | Phase IIa GlaxoSmithKline (GSK study KIP113049) (NCT01110057) | No significant alleviation in patients with lumbosacral radiculopathy |
| | | Phase II GlaxoSmithKline (GSK study Id: KIP112967) (NCT00969059) | No significant reduction in patients with peripheral nerve injury |
| **MINOCYCLINE** | Microglia (unclear molecular target) | Phase II University of Adelaide (Australia) | No significant on capsaicin challenge in patients with sciatica. Trend to decrease pain prior to challenge [18] |
| | | Phase II University of Adelaide (Australia) | Underway (Post-operative intercostal neuralgia) |
| **PENTOXIFYLLINE** | Developed as a PDE blocker, also acts on other targets. | Case report University of California Los Angeles (USA) | Improvement in diabetic neuropathy [30] |
| | | Phase II Yale University, Write State University, University of California Irvine (USA), | No benefit in diabetic neuropathy [31-33] |
| | | Phase II Jagiellonian University (Poland) | Reduced postoperative opioid requirements when administered preemptively [36] |

Table 1: contd….

| | | Phase II Jagiellonian University (Poland) | No effect on postoperative pain when given postoperatively [35] |
|---|---|---|---|
| | | Phase II National Defense University (Taipei, Rep China) | Reduced need for postoperative opiates when administered preemptively [34] |
| **PROPENTOFYLLINE** | Developed as a PDE blocker, also acts on other targets | Phase II Dartmouth Medical School (USA) | No effect on post-herpetic neuralgia [39] |
| **IBUDILAST (AV-411, MN166)** | Developed as a PDE blocker, also acts on other targets, especially MIF and TLR4 signaling | Phase I and II Avigen (NCT00576277) | Non-significant trend over placebo in patients with diabetic neuropathy and CRPS. |
| | | Phase I and II Aalborg University (Denmark) (NCT01389193) | Underway (Migraine) |
| **SULFASALAZINE** | Anti-bacterial and anti-inflammatory (multiple targets: NFKB, tetrahydrobiopterin synthesis) | Phase II Massachusetts General Hospital (USA) (NCT01667029) | Not yet recruiting. Will investigate effects on painful diabetic neuropathy |

## A SECOND LIFE FOR MEDICINES WITH GLIAL EFFECTS

In addition to the search for new microglial/astrocytic pathways and glia-targeting molecules, another strategy has been the use of already known compounds that present glia-modulatory effects. Several candidates have been identified in the available pharmacopeia, which display the ability to interfere with glial functions; some of them have been already routinely utilized for decades for other indications. For several of these molecules, clinical trials are already conducted or completed in chronic pain patients (Table **1**).

### Minocycline

The semi-synthetic second generation tetracycline minocycline has been used for more than three decades against bacterial infections, and acne. As previously mentioned, in experimental rodent models of chronic pain minocycline has demonstrated a remarkable efficacy at reducing microglial reaction and preventing pain behaviors. Such results strongly support the abundance of converging data that showed the regulatory influence of tetracyclines on many biochemical processes, including immune and inflammatory cells independently from their antimicrobial actions [16]. Despite the theoretical side effects of a complete microglial inhibition mentioned above, the good insight about the use of minocycline in patients over the last four decades prompted for clinical trials. Minocycline and to a lesser extent the other natural or semisynthetic tetracyclines are now considered plausible candidates for chronic pain treatments [9]. The pharmacokintetics properties of minocycline have been described in details in many studies [17]. Interestingly, minocycline has good absorption (above 95-100% after oral administration, with a $C_{max}$ reached by 2h), distribution (Vd = 80-115L, or 1.17L/kg) and some extent of blood-brain barrier penetration probably due to relative high lipid solubility as compared to other tetracyclines, sustaining its potential use in oral administration. Minocycline tolerability is satisfactory, but side effects have been reported to be above what is observed for other tetracyclines [18]; the reported undesirable effects associated with long-term minocycline in particular may include dizziness, gastrointestinal concerns, multiple organ hypersensitivity or modification of skin pigmentation.

The results of a randomized, double-blinded, placebo controlled and cross over clinical trial have been recently published that compared the effect of oral minocycline or pregabalin on pain in patients with chronic unilateral sciatica before or after a pronociceptive challenge induced by intradermal administration of the pungent component capsaicin [19]. Although the effects of minocycline were not statistically significant in this Phase IIa trial, it displayed a tendency to reduce allodynia and hyperalgesia, especially prior to capsaicin injection. The low number of patients included (eighteen) and the apparent weak effect of the capsaicin challenge reduce the significance of this study, but the small effect of minocycline might constitute a foretaste of an anti-nociceptive effect in humans. Further results might be released in a near future as another Phase II clinical trial is currently ongoing about the effect of minocycline on post-operative intercostal neuralgia (identification number NCT01314482).

## Ceftriaxone

Another antibiotic has gain visibility in neuropathology and especially in chronic pain. Ceftriaxone, a β-Lactam antibiotic from the cephalosporin sub-class (third generation), is capable of selectively increasing the expression of the astrocytic glutamate transporter EAAT-2. Ceftriaxone is well tolerated with minor and transient side effects. It displays several unusual pharmacokinetics hallmarks among cephalosporins, especially a long elimination half-life [20]. These properties make it a good candidate for many neurological diseases presumably associated with glutamatergic-based excitotoxicity such as epilepsy, amyotrophic lateral sclerosis or stroke. In particular, it has been successfully used in various animal models of chronic pain including diabetic neuropathy [21], visceral hypersensitivity [22, 23], peripheral nerve injury [24] and opioid-induced hyperalgesia [25]. Beyond the usual concern regarding the safety of a long-term use of antibiotics in patients, clinical trials are eagerly awaited to evaluate the analgesic capacity of ceftriaxone in chronic pain patients. However, considering ceftriaxone ability to reinforce EAAT-2 activity, great care should be given to potential neurologic or psychiatric side effects emerging from a widespread increased glutamate transport in the CNS.

## Pentoxifylline/Propentofylline

Pentoxifylline is a xanthine derivative mostly used to improve intermittent claudication in patients with obstructed arteries, but it has a large range of other applications from alcoholic/non-alcoholic steatohepatitis to Peyronie's disease. The wide spectrum of mechanisms of actions displayed by pentoxiphylline includes inhibition of phosphodiesterase, reduction of TNFα signaling and lessening of inflammation. Propentofylline, a structurally close relative, has been less employed in clinics but shares similar properties. Strikingly, among these biological actions, pentoxifylline and propentofylline were shown to be inhibitors of glial activation [26, 27] and effective at reducing signs of hypersensitivity in animal models of neuropathic pain [28-30]. The idea of treating chronic neuropathic pain with pentoxifylline has first emerged it the late 1980's for diabetic neuropathy [31] but these results were not confirmed by further studies [32-34]. Nonetheless, although needing confirmation, it was reported that pentoxifylline was effective at reducing post-operative pain in patients when administered preemptively but not postoperatively [35-37]. Similarly, propentofylline exhibits antiallodynic properties in models of neuropathic pain [38] but negative results, albeit debated [39], were obtained in patients affected with post-herpetic neuralgia [40].

## Ibudilast (AV-411; MN-166; KC-404)

Ibudilast (3-isobutyryl-2-isopropylpyrazolo[1,5-a]pyridine) is an orally bioavailable vaso- and broncho-dilatator, developed as a phosphodiesterase inhibitor, which has been used for two decades against asthma and stroke [41]. Nevertheless, its spectrum of action is widely non-specific including inhibition of macrophage migration inhibitory factor (MIF), reduction of cytokine release and platelet aggregation, Toll Like Receptor 4 antagonism and a reduction of glial reaction [42]. Ibudilast crosses the blood brain barrier making it a good candidate for neurological diseases and especially neuropathic pain.

In rodents, ibudilast has shown efficacy at reducing hypersensitivity in different experimental models [43-45]. Chronic pain-oriented Phase I-II trial in humans (clinical trial identifier NCT00576277) has shown that single or multiple regimens of ibudilast are relatively well tolerated at exposure doses comparable to rodent studies [46]. In addition, this trial reveals a non-significant trend of pain improvement over placebo in patients with diabetic neuropathy and complex regional pain syndrome (CRPS), prompting for more advanced studies. Another double blind, randomized and placebo controlled clinical trial is currently investigating the effect of ibudilast on migraine (identifier NCT01389193).

The amino conjugate of ibudilast, AV1013, seems to present glial inhibitory properties without retaining phosphodiesterase-blocking capacity [47]. AV1013 might therefore become an interesting future candidate for chronic pain. But, to the author's knowledge, no experimental data has been released in that direction so far, neither has any clinical trial been started to evaluate this possibility.

## Sulfasalazine

Sulfasalazine is a disease-modifying drug chiefly used in inflammatory bowel diseases (namely Crohn's disease and ulcerative colitis) as well as joint diseases, especially rheumatoid arthritis. It is an azo-bound sulfapyridine-mesalamine (sulfonamide and aminosalicylate) conjugate, cleaved into both compounds by intestinal bacteria. The mechanisms of action, although incompletely understood, may stem from biological effects of either sulfasalazine itself, sulfapyridine, mesalamine or a combination of the three molecules and target various pathways including nuclear factor kappa B (NFκB), cystine transport, tetrahydrobiopterin synthesis and NO production [48, 49]. The current clinical use of sulfasalazine is based on an assumed dominance of its peripheral anti-inflammatory actions, which is further supported by the poor absorption and bioavailability of the compound or its metabolites (30-60%). However sulfasalazine has been claimed to cross the blood-brain barrier, making CNS cells potential targets. In this context, convergent lines of evidence indicate the ability of sulfasalazine to interfere with activations of cultured astrocytes and microglia induced by bacterial lipopolysaccharide [50, 51].

The prospective use of sulfasalazine in pain management has recently crystalized attention. A recent meta-analysis has unveiled the significantly higher efficacy of sulfasalazine over other disease-modifying drugs at reducing pain in several forms of inflammatory arthritis. More striking was the report of a reduction of mechanical hypersensitivity in a rat model of diabetic neuropathy induced by orally administered sulfasalazine [52]. This therefore opens new perspectives in the use of sulfasalazine in neuropathic pain and a recent Phase II clinical trial exploring the efficacy of sulfasalazine in painful neuropathy has been recently launched (clinical trial identifier NCT01667029).

## SUMMARY, CONCLUDING REMARKS AND CAVEATS

Astrocytes and microglia are now deemed plausible targets for new pharmacological developments against chronic pain. However, the full inhibition of glial function is not a realistic option, prompting for more elaborated strategies aiming to disrupt specific glial pathways important in chronic pain. The evaluation of p38 MAPK inhibition in patients seems promising but further trials are needed. Besides, several known compounds display the ability to curb glial activation or modulate glial physiology and are currently tested in clinical trials. The well documented pharmacokinetics and pattern of biological effects of these medicines strongly justify their evaluation in additional clinical trials and make them credible candidates as possible adjuvants options in addition to usual neuron-oriented treatments.

Importantly however, the systemic route of administration of all these molecules cannot fully ensure the CNS (spinal) site of modulation, as peripheral inflammatory or nerve target can possibly be affected as well, nor can it ensure the accountability of glial inhibition in symptom relief. Furthermore, a critical reflection should be made about the rationale of the long-term utilization antibiotics such as minocycline or ceftriaxone considering their impact on the commensal microbiotic flora. In addition, specific safety concerns suggest likely limiting long-term use of minocycline due to well-known side effect profile, and to give foremost attention to ceftriaxone neurological side effects. However, these molecules might constitute the starting point for the synthesis of closely related new products, which retain the ability to limit glial physiology with reduced spectrum of undesirable effects.

## ACKNOWLEDGEMENTS

Declared none.

## CONFLICT OF INTEREST

The authors confirm that this chapter content has no conflict of interest.

## REFERENCES

[1]     Astyrakaki E, Papaioannou A, Askitopoulou H. References to anesthesia, pain, and analgesia in the Hippocratic Collection. Anesth Analg 2010; 110(1): 188-94.

[2]     Vane JR. The fight against rheumatism: from willow bark to COX-1 sparing drugs. J Physiol Pharmacol 2000; 51(4 Pt 1): 573-86.

[3]     Vo T, Rice AS, Dworkin RH. Non-steroidal anti-inflammatory drugs for neuropathic pain: how do we explain continued widespread use? Pain 2009; 143(3): 169-71.

[4]     Gosselin RD, Suter MR, Ji RR, Decosterd I. Glial Cells and Chronic Pain. Neuroscientist 2010; 16(5): 519-31.

[5]     Kettenmann H, Hanisch UK, Noda M, Verkhratsky A. Physiology of microglia. Physiol Rev 2011; 91(2): 461-553.

[6]     Bennarroch EE. Central neuron-glia interactions and neuropathic pain: overview of recent concepts and clinical implications. Neurology 2010; 75(3): 273-8.

[7]    Meller ST, Dykstra C, Grzybycki D, Murphy S, Gebhart GF. The possible role of glia in nociceptive processing and hyperalgesia in the spinal cord of the rat. Neuropharmacology1994; 33(11): 1471-8.

[8]    Proudfoot AT, Bradberry SM, Vale JA. Sodium fluoroacetate poisoning. Toxicol Rev 2006; 25(4): 213-9.

[9]    Bastos LF, de Oliveira AC, Watkins LR, Moraes MF, Coelho MM. Tetracyclines and pain. Naunyn Schmiedebergs Arch Pharmacol 2012; 385(3): 225-41.

[10]   Ji RR, Gereau RWt, Malcangio M, Strichartz GR. MAP kinase and pain. Brain Res Rev 2009; 60(1): 135-48.

[11]   Svensson CI, Hua XY, Protter AA, Powell HC, Yaksh TL. Spinal p38 MAP kinase is necessary for NMDA-induced spinal PGE(2) release and thermal hyperalgesia. Neuroreport 2003; 14(8): 1153-7.

[12]   Svensson CI, Marsala M, Westerlund A, *et al.* Activation of p38 mitogen-activated protein kinase in spinal microglia is a critical link in inflammation-induced spinal pain processing. J Neurochem 2003; 86(6): 1534-44.

[13]   Tsuda M, Mizokoshi A, Shigemoto-Mogami Y, Koizumi S, Inoue K. Activation of p38 mitogen-activated protein kinase in spinal hyperactive microglia contributes to pain hypersensitivity following peripheral nerve injury. Glia 2004; 45(1): 89-95.

[14]   Sorkin L, Svensson CI, Jones-Cordero TL, Hefferan MP, Campana WM. Spinal p38 mitogen-activated protein kinase mediates allodynia induced by first-degree burn in the rat. J Neurosci Res 2009; 87(4): 948-55.

[15]   Anand P, Shenoy R, Palmer JE, *et al.* Clinical trial of the p38 MAP kinase inhibitor dilmapimod in neuropathic pain following nerve injury. Eur J Pain 2011; 15(10): 1040-8.

[16]   Griffin MO, Fricovsky E, Ceballos G, Villarreal F. Tetracyclines: a pleitropic family of compounds with promising therapeutic properties. Review of the literature. Am J Physiol Cell Physiol 2010; 299(3): C539-48.

[17]   Agwuh KN, MacGowan A. Pharmacokinetics and pharmacodynamics of the tetracyclines including glycylcyclines. J Antimicrob Chemother 2006; 58(2): 256-65.

[18]   Ochsendorf F. Minocycline in acne vulgaris: benefits and risks. Am J Clin Dermatol 2010; 11(5): 327-41.

[19]   Sumracki NM, Hutchinson MR, Gentgall M, Briggs N, Williams DB, Rolan P. The effects of pregabalin and the glial attenuator minocycline on the response to intradermal capsaicin in patients with unilateral sciatica. PLoS One 2012; 7(6): e38525.

[20]   Bijie H, Kulpradist S, Manalaysay M, Soebandrio A. *In vitro* activity, pharmacokinetics, clinical efficacy, safety and pharmacoeconomics of ceftriaxone compared with third and fourth generation cephalosporins: review. J Chemother 2005; 17(1): 3-24.

[21]   Gunduz O, Oltulu C, Buldum D, Guven R, Ulugol A. Anti-allodynic and anti-hyperalgesic effects of ceftriaxone in streptozocin-induced diabetic rats. Neurosci Lett 2011; 491(1): 23-5.

[22]   Yang M, Roman K, Chen DF, Wang ZG, Lin Y, Stephens RL, Jr. GLT-1 overexpression attenuates bladder nociception and local/cross-organ sensitization of bladder nociception. Am J Physiol Renal Physiol 2011; 300(6): F1353-9.

[23]   Lin Y, Roman K, Foust KD, Kaspar BK, Bailey MT, Stephens RL. Glutamate Transporter GLT-1 Upregulation Attenuates Visceral Nociception and Hyperalgesia *via* Spinal Mechanisms Not Related to Anti-Inflammatory or Probiotic Effects. Pain Res Treat 2011; 2011: 507029.

[24]    Hu Y, Li W, Lu L, Cai J, Xian X, Zhang M, *et al.* An anti-nociceptive role for ceftriaxone in chronic neuropathic pain in rats. Pain 2010; 148(2): 284-301.

[25]    Chen Z, He Y, Wang ZJ. The beta-lactam antibiotic, ceftriaxone, inhibits the development of opioid-induced hyperalgesia in mice. Neurosci Lett 2012; 509(2): 69-71.

[26]    Chao CC, Hu S, Close K, *et al.* Cytokine release from microglia: differential inhibition by pentoxifylline and dexamethasone. J Infect Dis 1992; 166(4): 847-53.

[27]    Schwartz D, Engelhard D, Gallily R, Matoth I, Brenner T. Glial cells production of inflammatory mediators induced by Streptococcus pneumoniae: inhibition by pentoxifylline, low-molecular-weight heparin and dexamethasone. J Neurol Sci 1998; 155(1): 13-22.

[28]    Liu J, Feng X, Yu M, *et al.* Pentoxifylline attenuates the development of hyperalgesia in a rat model of neuropathic pain. Neurosci Lett 2007; 412(3): 268-72.

[29]    Liu J, Li W, Zhu J, *et al.* The effect of pentoxifylline on existing hypersensitivity in a rat model of neuropathy. Anesth Analg 2008; 106(2): 650-3.

[30]    Mika J, Osikowicz M, Makuch W, Przewlocka B. Minocycline and pentoxifylline attenuate allodynia and hyperalgesia and potentiate the effects of morphine in rat and mouse models of neuropathic pain. Eur J Pharmacol 2007; 560(2-3): 142-9.

[31]    Kalmansohn RB, Kalmansohn RW, Markham CH, Schiff DL. Treatment of diabetic neuropathy with pentoxifylline: case report. Angiology 1988; 39(4): 371-4.

[32]    Cohen KL, Lucibello FE, Chomiak M. Lack of effect of clonidine and pentoxifylline in short-term therapy of diabetic peripheral neuropathy. Diabetes Care 1990; 13(10): 1074-7.

[33]    Cohen SM, Mathews T. Pentoxifylline in the treatment of distal diabetic neuropathy. Angiology 1991; 42(9): 741-6.

[34]    Lee Y, Robinson M, Wong N, Chan E, Charles MA. The effect of pentoxifylline on current perception thresholds in patients with diabetic sensory neuropathy. J Diabetes Complications 1997; 11(5): 274-8.

[35]    Lu CH, Chao PC, Borel CO, *et al.* Preincisional intravenous pentoxifylline attenuating perioperative cytokine response, reducing morphine consumption, and improving recovery of bowel function in patients undergoing colorectal cancer surgery. Anesth Analg 2004; 99(5): 1465-71; table of contents.

[36]    Szczepanik AM, Wordliczek J, Serednicki W, Siedlar M, Czupryna A. Pentoxifylline does not affect nociception if administered postoperatively. Pol J Pharmacol 2004; 56(5): 611-6.

[37]    Wordliczek J, Szczepanik AM, Banach M, *et al.* The effect of pentoxifiline on post-injury hyperalgesia in rats and postoperative pain in patients. Life Sci 2000; 66(12): 1155-64.

[38]    Sweitzer S, De Leo J. Propentofylline: glial modulation, neuroprotection, and alleviation of chronic pain. Handb Exp Pharmacol 2011(200): 235-50.

[39]    Watkins LR, Hutchinson MR, Johnson KW. Commentary on Landry *et al.*: "Propentofylline, a CNS glial modulator, does not decrease pain in post-herpetic neuralgia patients: *in vitro* evidence for differential responses in human and rodent microglia and macrophages". Exp Neurol 2012; 234(2): 351-3.

[40]    Landry RP, Jacobs VL, Romero-Sandoval EA, DeLeo JA. Propentofylline, a CNS glial modulator does not decrease pain in post-herpetic neuralgia patients: *in vitro* evidence for differential responses in human and rodent microglia and macrophages. Exp Neurol 2012; 234(2): 340-50.

[41]    Kishi Y, Ohta S, Kasuya N, Sakita S, Ashikaga T, Isobe M. Ibudilast: a non-selective PDE inhibitor with multiple actions on blood cells and the vascular wall. Cardiovasc Drug Rev 2001; 19(3): 215-25.

[42]    Rolan P, Hutchinson M, Johnson K. Ibudilast: a review of its pharmacology, efficacy and safety in respiratory and neurological disease. Exp Opin Pharmacother 2009; 10(17): 2897-904.

[43]    Hama AT, Broadhead A, Lorrain DS, Sagen J. The antinociceptive effect of the asthma drug ibudilast in rat models of peripheral and central neuropathic pain. J Neurotrauma 2012; 29(3): 600-10.

[44]    Ledeboer A, Liu T, Shumilla JA, *et al.* The glial modulatory drug AV411 attenuates mechanical allodynia in rat models of neuropathic pain. Neur Glia Biol 2006; 2(4): 279-91.

[45]    Lilius TO, Rauhala PV, Kambur O, Kalso EA. Modulation of morphine-induced antinociception in acute and chronic opioid treatment by ibudilast. Anesthesiology 2009; 111(6): 1356-64.

[46]    Rolan P, Gibbons JA, He L, *et al.* Ibudilast in healthy volunteers: safety, tolerability and pharmacokinetics with single and multiple doses. Br J Clin Pharmacol 2008; 66(6): 792-801.

[47]    Snider SE, Vunck SA, van den Oord EJ, Adkins DE, McClay JL, Beardsley PM. The glial cell modulators, ibudilast and its amino analog, AV1013, attenuate methamphetamine locomotor activity and its sensitization in mice. Eur J Pharmacol 2012; 679(1-3): 75-80.

[48]    Wahl C, Liptay S, Adler G, Schmid RM. Sulfasalazine: a potent and specific inhibitor of nuclear factor kappa B. J Clin Invest 1998; 101(5): 1163-74.

[49]    Chidley C, Haruki H, Pedersen MG, Muller E, Johnsson K. A yeast-based screen reveals that sulfasalazine inhibits tetrahydrobiopterin biosynthesis. Nat Chem Biol 2011; 7(6): 375-83.

[50]    Lamirand A, Ramauge M, Pierre M, Courtin F. Bacterial lipopolysaccharide induces type 2 deiodinase in cultured rat astrocytes. J Endocrinol 2011; 208(2): 183-92.

[51]    Wilms H, Rosenstiel P, Sievers J, Deuschl G, Zecca L, Lucius R. Activation of microglia by human neuromelanin is NF-KB-dependent and involves p38 mitogen-activated protein kinase: implications for Parkinson's disease. FASEB J 2003; 17(3): 500-2.

[52]    Berti-Mattera LN, Kern TS, Siegel RE, Nernet I, Mitchell R. Sulfasalazine blocks the development of tactile allodynia in diabetic rats. Diabetes 2008; 57(10): 2801-8.

Send Orders for Reprints to reprints@benthamscience.net

# CHAPTER 3

## Spinal Cord Injury: Clinical Trials and Translational Tribulations

**Simon Harris, Jared Wilcox and Michael G. Fehlings**[*]

*Divisions of Orthopaedic Surgery and Neurosurgery, University of Toronto, Canada and the Division of Genetics and Development, Toronto Western Research Institute, Toronto, Canada*

**Abstract:** Spinal Cord Injury (SCI) is a common cause of neurological morbidity and mortality, particularly affecting young adults. The effect on an individual and his or her family can be devastating. The prevalence of SCI in Canada and the US is estimated at 1.4 million cases with an annual incidence of 13,000. The Pathophysiology of SCI is a complex interaction of inflammation, vascular insult, glial scarring and cell death, dysfunction and regeneration. Despite a growing wealth of pre-clinical scientific knowledge and understanding, there has been only minimal clinical benefit shown in a few human trials. In this chapter we provide an in-depth review of clinical trials in SCI identifying concurrent novel, on-going trials and highlight the hopes for the future. The chapter will focus on the key historical trials of methylprednisolone, GM-1 and thyrotrophin-releasing hormone and then discuss the potential of ion-channel blockers, targeted therapeutics and monoclonal antibodies. Further discussion focuses on endocrine and immunomodulation and injectable biomaterials. The experiences of human SCI stem cell trials are analysed and the novel therapeutic target of central pattern gait generator stimulation in neuro-rehabilitation is proposed.

**Keywords:** Spinal cord injury, clinical trials, therapeutic targets, NASCIS, hydrogels, self-assembling peptides, HAMC, stem cells, Geron, timing of surgery, immunomodulation, prevention, GM-1, thyrotrophin-releasing hormone.

## INTRODUCTION

Spinal Cord Injury (SCI) has devastated the lives of millions of individuals around the world. Surgical and peri-operative care continues to advance [1] and we have seen a growing wealth of scientific understanding of the pathophysiological mechanisms underlying SCI over the past twenty years [2-4].

---

[*]**Address correspondence to Michael G. Fehlings:** Divisions of Orthopaedic Surgery and Neurosurgery, University of Toronto, Canada and the Division of Genetics and Development, Toronto Western Research Institute, Toronto, Canada; Tel: 1-416-603-5627; Fax: 1-416-603-5298; E-mail: michael.fehlings@uhn.ca

**Atta-ur-Rahman (Ed)**
**All rights reserved-© 2013 Bentham Science Publishers**

Despite these advances, we are unable to fully reverse neurological injury in SCI and there have been few treatments that have jumped the translational gap from rodent model to standard care of injured humans.

Basic science research in animal models of SCI has elucidated many potential therapeutic targets utilising both non-invasive [5], invasive [6] and cellular strategies [7]. The hope is that one of these targets, or a combination of many, will prove successful in human SCI. The reality is that the Holy Grail of SCI treatment is still a dream for the future, but a dream worth chasing. This chapter will provide a detailed overview of completed and ongoing human trials in SCI and potential avenues for future research.

## SURGICAL CARE OF THE SPINAL CORD INJURED PATIENT

### The Edwin Smith Papyrus

The first documented report on acute SCI can be dated back to the Egyptians in 2500BC [8]. The translation from the original papyrus, performed by Professor Edwin Smith, describes the examination of a patient that had a fall from height onto his head (case 33):

> "Thou findest that one vertebra has fallen into the next one, while he is voiceless and cannot speak… thou find he is unconscious of his two arms and his two legs" [8].

This was indeed considered an "ailment not to be treated" [8]. The concept of acute SCI being an untreatable affliction prevailed for many millennia. The advent of sterile technique, surgical spinal decompression and spinal instrumentation has proved this dogma false. Today patients with acute SCI are surviving traumas that they may not have survived only a few decades earlier. No single intervention, technique or care pathway has been the keystone to achieving this success. Indeed, there are multiple factors which have incrementally influenced the current state of affairs in SCI treatment [9]. Firstly, integrated, well-planned trauma systems have expedited the speed of transfer of patients with SCI to hospitals equipped to deal with spine surgery. Patients with multi-organ injuries are being triaged and treated appropriately. Peri-operative care of the SCI patient has evolved into a multi-disciplinary team approach in specialised tertiary referral centres [10]. The final step in recovery is rehabilitation and this too has seen vast improvements [11-15].

Annual total incidence of SCI ranges from 12 to 71 per million population [16-19]. The true incidence is likely higher due to pre-hospital mortality exclusion [20]. The epidemiology of acute SCI is such that it is often considered as primarily a disease of young males. Fast driving and risk taking behaviours put this group at high risk. However, recent data from the US has identified a trend of increasing age of acute SCI [21]. Unfortunately the survival rate of SCI in the elderly population remains poor and thus the prevalence of elderly patients with SCI is unlikely to increase [21].

A 10 year retrospective cohort analysis of SCI performed at the University of British Columbia, which covers a population of 4 million people, found an incidence of 35.7 per million with a median age of 34.5-45.5 years and a male to female ratio of 4.4:1. Interestingly, despite an increasing rate of surgical intervention (61.8% to 86.4%) there was no improvement in mortality rate or length of hospital stay. SCI patients over 75 years old had a 20% mortality rate [22].

## Classification of Spinal Cord Injury

In order to classify the constellation of deficits following SCI – for clinical or research purposes – a standardized assessment tool was developed. The American Spinal Injury Association (ASIA) impairment scale was developed which describes these injury categories:

- **A – Complete:** No motor or sensory function is preserved in the sacral segments S4 or S5.

- **B – Incomplete:** Sensory but not motor function is preserved in the sacral segments S4 or S5.

- **C – Incomplete:** Motor function is preserved below the level of injury, with power less than grade 3.

- **D – Incomplete:** Motor function is preserved below the level of injury, with power greater than grade 3.

- **E – Normal:** No motor or sensory deficit.

There are 5 commonly recognized clinical syndromes of spinal cord injury, which are determined by the pathological insult and cross sectional anatomy at the level of the injury:

1. **Central Cord Syndrome** – Upper limb impairment greater than lower limb. Commonly occurs in cervical spine hyperextension injuries with underlying cervical spine stenosis.

2. **Brown-Sequard Syndrome (Hemisection of the cord)** – Loss of ipsilateral motor function, fine touch and vibration sense and loss of contralateral pain and temperature sensation.

3. **Anterior Cord Syndrome** – Compromise of the anterior spinal vascular supply resulting in loss of motor power with preservation of fine touch and proprioception.

4. **Conus Medullaris Syndrome** – Compression at level of the conus results in a mixed upper and lower motor neuron pattern distal to the lesion.

5. **Cauda Equina** – Bilateral leg weakness, faecal incontinence, urinary retention and perineal saddle anaesthesia due to compression of the lower motor neurons within the distal spinal canal.

In addition to the severity of sensorimotor deficit, the spinal level of injury determines the patient's functional outcome. For example, in cervical SCI, preservation of C6 allows the patients to feed themselves *via* elbow flexion, whilst C7 function allows patients to propel themselves in a wheelchair due to the presence of triceps motor power. This highlights the importance of C7 for patient independence. For all SCI, approximately 34.3% suffered incomplete tetraplegia, 25.1% complete paraplegia, 22.1% complete tetraplegia and 17.5% incomplete paraplegia.

**Timing of Surgical Intervention**

The evidence justifying early surgical decompression in cervical SCI patients is increasing in volume and quality. In 1999, Mirza *et al.* showed that patients suffering from acute cervical SCI had improved neurological recovery and shorter hospitalization if operated within 72 hours when compared to a cohort that underwent closed reduction followed by observation and subsequent surgery within 10 to 14 days [23]. In 2012, Fehlings *et al.* as part of the Surgical Timing in Acute Spinal Cord Injury Study (STASCIS), published the strongest evidence to date in favor of early surgical intervention. As part of a multicenter, international, prospective study of cervical SCI, the group showed that 19.8% of patients undergoing early surgery (mean 14.2 hours from time of injury) compared to 8.8% in the late decompression group (mean 48.3 hours), improved their ASIA impairment scale by ≥ 2 grades at 6 months [24]. The use of the ASIA impairment scale in clinical trials such as this is limited by the ceiling effect. For example, a patient with an ASIA D impairment can only improve to an ASIA E and therefore can not improve by 2 or more grades.

There is equipoise in the evidence for timing of surgery in thoraco-lumbar SCI with respect to functional improvements. However, early surgery for all levels and severity of SCI has additional morbidity improvements including a reduction in ventilator-dependent days, hospital stay, total in-hospital cost, pneumonia, pressure sores and urinary tract infections [25-29].

Timing of surgery is not, however, the most important factor in predicting clinical outcome. A systematic review of 51 published clinical trials in 2012 identified the severity of neurological injury level of injury, and the presence of a zone of partial preservation as consistent predictors of neurological outcome [30].

## GUIDELINES FOR CONDUCTING HUMAN TRIALS

### Phases of Human Clinical Trials

There stands a large translational gap between interventions that have shown promise in rodent models of acute SCI and those that have become proven therapies for human patients. One of the key tools for traversing this inter-species knowledge canyon is the use of carefully planned, safe and intensely monitored human trials. Clinical trials are organized into four phases (Table **1**).

**Table 1:** The phases of human clinical trials [31]

| Phase | Purpose |
|---|---|
| 1 | Screening for safety: Study drug is tested in a small group of patients (50) to evaluate its safety, identify side effects and dose range. |
| 2 | Establishing testing protocol: Study drug is tested on a larger number of patients (200) to evaluate its effectiveness and further test its safety. |
| 3 | Final Testing: Study drug is tested on a large number of patients (2000) to confirm its effectiveness in comparison to other commonly used treatments and to further monitor its safety. |
| 4 | Post-approval studies: Delineate additional information including drugs risks, benefits and side effects. |

## The International Campaign for Cures of Spinal Cord Injury Paralysis

In conducting clinical trials, patient safety is paramount. The International Campaign for Cures of Spinal Cord Injury Paralysis (ICCP) produced a series of guidelines on how to safely design clinical trials [32], identify consistent outcome measures [33], and identify suitable patient populations [34, 35]. With strict adherence to such guidelines, the quality of the data generated will be consistent and superior to studies conducted without these guidelines.

## Global Human Clinical Trials

The Ottawa Statement and the Declaration of Helsinki delineate guidelines on the safe registration, operation and reporting of global human clinical trials with a core focus on transparency [36]. These guidelines have been well received by the World Health Organization, Canadian Institute for Health Research and the US National Institute for Health [37, 38]. Many countries have begun legislation in this new phase of medical research with human participants, such as the Fair Access to Clinical Trials Act in the United States.

## Controlled and Uncontrolled Trials

Controlled trials represent the direct contemporaneous experimental comparison of one intervention with another. Uncontrolled trials involve the testing of an intervention without a concurrent untreated or gold-standard intervention group. Controlled trials are subject to multiple level watchdog monitoring and strict legal governance. Uncontrolled trials, however, are not governed by such strict legislations or regulations and thus may operate below the exemplary standards upheld by controlled trials. Industry and public institutions have slowly conformed to these standards and transparencies [38-40]. However, the data reported in such registries is often incomplete, changed or inaccurate [41]. Clear, well maintained registries, accurate reporting and transparency of outcomes – good or bad – is essential to regain public trust and the financial and legal support of government.

## THERAPEUTIC TARGETS AND THE PATHOPHYSIOLOGY OF SPINAL CORD INJURY

### Primary and Secondary Injury

Acute traumatic injury to the spinal cord can come in the form of a blunt mechanism, such as an acrobat falling onto a hard surface or a person thrown from a car during an automobile collision, or a penetrating injury such as a stab or gunshot wound. The forces involved in each of these events are transmitted to the spinal column and, if great enough, result in disruption of the bony and ligamentous structures and result in damage to the neural elements [42]. Damage to the spinal cord or the exiting nerve roots can result in motor, sensory or autonomic dysfunction [43]. In an attempt to delineate the precise cause of neurological dysfunction, researchers have divided the temporal sequence of destructive events into primary and secondary injury (Fig. **1**). Primary injury refers to the destructive forces that directly damage the neural structures such as the shear force tearing an axon or a direct compressive force occluding a blood vessel resulting in ischemia [44]. These destructive primary mechanisms not only result in instantaneous damage to neurons and blood vessels but also initiate a cascade of cellular mechanisms that result in ongoing damage to the neural structures, termed secondary injury [43, 45, 46]. In fact, in cases of ongoing primary injury, for example in the setting of a fracture dislocation where the bony spinal column is displaced and physically pushed against the spinal cord, these cellular mechanisms are thought to be locked into the 'on' position until such physical forces are removed either by closed reduction or surgically. Autonomic dysregulation in particular causes decreased cardiac output and reduced vascular tone contributing to hypotension, which may in turn induce critical ischemia of the spinal cord in the peri-operative period [47, 48]. Secondary injury may persist from hours to weeks to years following primary injury [43]. Pathophysiological events contributing to SCI include lipid peroxidation, eicosanoid and prostaglandin formation, protease activation, free radical production and excitotoxic molecule release increasing intracellular calcium [2, 43]. This produces significant inflammatory changes in the spinal cord [49]. A great deal of work has gone into the detailed understanding of these cellular cascades and along with this work has come an appreciation for the destructive effects of the mechanisms and the role of potential therapeutics to halt these events.

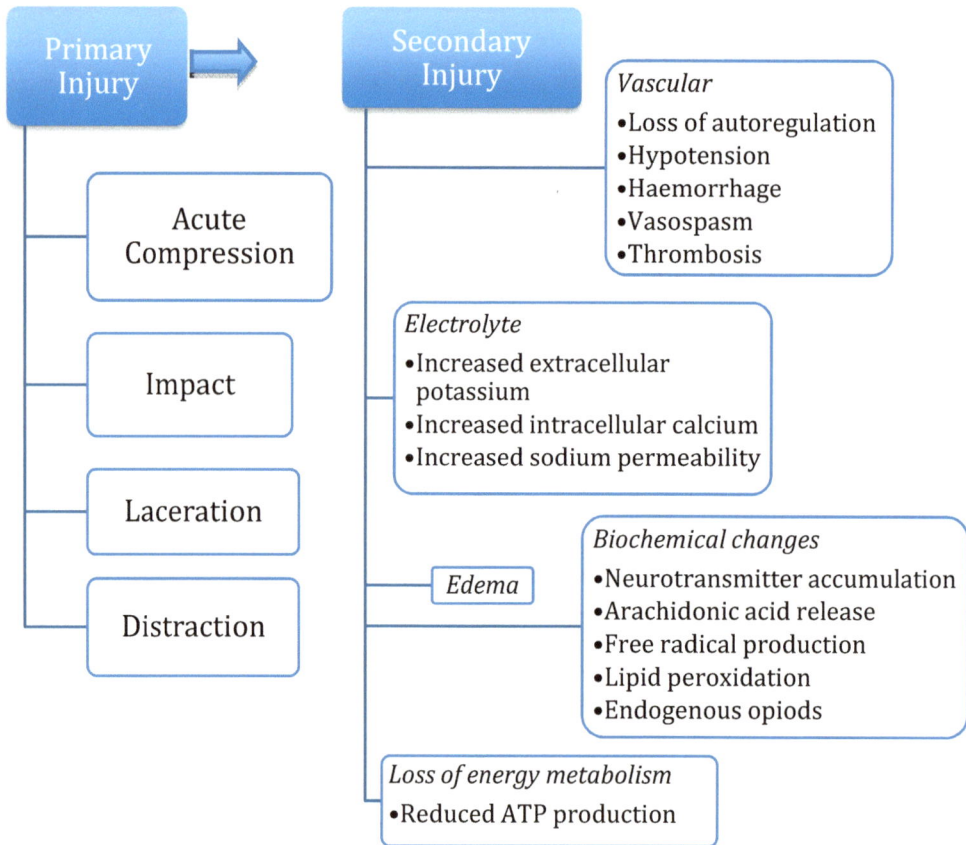

**Figure 1:** Primary and secondary injury mechanisms.

## Blood-Spinal Cord Barrier

The Blood-Spinal Cord Barrier (BSCB) is a unique microenvironment providing specialised support for the cells and their functions within the spinal cord. The function of the BSCB is based upon non-fenestrated endothelial cells, basement membrane, pericytes and astrocyte end feet processes [50]. In the event of traumatic SCI, the BSCB is disrupted at a cellular and macrovascular level. This in turn causes further damage but may also provide an opportunity. In the injured state, capillaries are physically disrupted allowing chemical or cellular therapies injected systemically to be able to traverse the BSCB and directly affect the injured axons and supporting cells [51, 52]. Complex secondary pathophysiological mechanisms cause extension of damage within the spinal cord parenchyma [51, 52].

Disruption of the function of the BSCB occurs within minutes of injury [53]. The re-establishment of the BSCB occurs within 14-28 days post-injury [51, 54] when analysed by traceable markers crossing from the systemic blood circulation into the spinal cord tissue. When analysed by Dynamic Contrast-Enhanced MRI, the BSCB remains functionally compromised for up to 56 days post-injury [55].

## PREVENTION IS BETTER THAN A CURE

### Primary Prevention of SCI

The focus of this chapter is the treatment of acute SCI. However, no discussion on acute SCI management is complete without re-iterating the importance of preventing the initial occurrence. The main mechanisms of injury are motor vehicle collisions, falls from height and recreational activities. The incidence of motor vehicle collisions can be significantly reduced by abiding to local speed limits whilst maintaining a safe distance between vehicles and over taking with caution. Providing alternate means of returning home when alcohol or recreational drugs have been consumed is essential. The majority of falls are work place injuries. Strict adherence to work place safety guidelines is paramount. Enjoyment with caution and within the individual's physical limits of control is important during recreational activities such as skiing and all-terrain vehicle use.

The common thread within all primary prevention for musculoskeletal injuries is education and adherence to advice and guidelines. The risk factors for SCI are well-known and largely avoidable and yet people continue to engage in risky behaviour. For this reason SCI incidence is likely to remain stable and its prevalence in society will continue to expand.

It is likely that true primary prevention can only be provided by novel innovation and removing the decision making from humans. For example the driverless car recently invented and being tested by Google may one day provide the technology to allow us all to travel safely. To be able to eradicate SCI is perhaps the goal that we should be striving for.

## Secondary Prevention of SCI

Secondary prevention is the act of minimising the severity of a disease once it has been diagnosed. In relation to acute SCI, this involves early diagnosis, initiation of spinal precautions (rigid cervical spine collar and hard back board) and transfer to spine centre for urgent surgical decompression and stabilisation of the injured spinal column. Early closed reduction of cervical spine fracture-dislocations is recommended [56]. Further, secondary prevention must include avoidance of hypotension (mean arterial pressure greater than 85mmHg) for 7 days following injury to maximise spinal cord perfusion [56, 57]. The patient should be cared for in a monitored setting to allow for accurate real-time invasive blood pressure monitoring [56].

## KEY HISTORICAL TRIALS IN ACUTE SCI MANAGEMENT

Clinical trials using pharmacological interventions to treat acute SCI have been performed for over thirty years [58]. These trials have mainly focused on systemically delivered medication for immunomodulation, membrane channel inhibition or micro-environment modification. A handful of drugs have been shown to be safe however there is debate as to whether there is strong enough evidence of benefit for these medications to become standard of care. This section focuses on three key historical drug trials; Methylprednisolone, monosialotetrahexosylganglioside (GM-1) and thyrotrophin-releasing hormone (TRH).

### Methylprednisolone and the NASCIS Trials

Methylprednisolone is an immunosuppressive corticosteroid administered as methylprednisolone sodium succinate (MPSS). The National Acute Spinal Cord Injury Study II (NASCIS-II) was a blinded randomized controlled study comparing MPSS against naloxone or placebo in 427 acute SCI patients [59]. There was no significant difference in neurological recovery at one year except on sub-group analysis of those patients receiving MPSS within 8 hours of SCI event. A subsequent trial (NASCIS-III) compared the use of MPSS for 24 or 48 hours post injury against tirilazad mesylate (a lipid peroxidation inhibitor) for patients that had suffered an acute SCI and had received an initial bolus of MPSS. There was no functional difference in those patients that received MPSS bolus within 3 hours of SCI event. If patients received MPSS bolus within 3 to 8 hours of injury, patients who received a 48-hour infusion of MPSS had a better motor recovery (but not functional improvement) at the cost of increased episodes of sepsis and pneumonia [60].

When critically appraised, the landmark findings of the NASCIS trials were criticised due to the small affected subgroup, failure to reach primary endpoints, inconsistency between centres, infection rates and blunt outcome measures employed [61, 62]. Despite a quarter of a century of research and the documented risks and benefits from the use of MPSS, its role in the management of acute SCI will vary from hospital to hospital and surgeon to surgeon and is declining in popularity [63, 64]. Hadley and Walters produced an evidence based summary article published in Neurosurgery (2013) evaluating the management steps in acute SCI, and state that the administration of MPSS is not recommended and the drug is not FDA approved for this role [56].

## Monosialotetrahexosylganglioside (GM-1) and the Largest Acute SCI Trial To Date

GM-1 (Sygen®, Fidia Pharmaceutical Corporation) is a ganglioside containing one sialic acid residue and has important broad roles in neuronal repair, plasticity and neutrophin release – potentially key factors in recovery from SCI. Published in 1991, Geisler *et al.* performed a prospective randomized placebo-controlled double blind trial of intravenous GM-1 in patients with acute SCI [65]. This small trial of 37 patients detected a significant difference between treatment arms with respect to neurological recovery, without complications and suggested a larger trial be conducted. The results were, however, heavily criticised by Schönhöfer [66] and separately by Landi and Ciccone [67] in correspondence with the New

England Journal of Medicine. The concerns focused on an uneven distribution of injury severity, a failure to show improvement in the American Spinal Injury Association (ASIA) score at two months and a failure to discuss or declare any side effects – the more serious association with gangliosides being the development of an ascending neuropathy (Guillain-Barré syndrome). A similar drug had previously been withdrawn from the German market in 1989 for the latter reason.

A large multicenter, double blind, randomised controlled trial of two doses of Sygen® *vs.* placebo (Sygen® Study Group) was then performed, recruiting 760 patients at 28 centres across North America to determine the drug's efficacy and safety. The ASIA scores showed a trend towards favourable outcomes with Sygen® but the study failed to prove significance [68-70]. Following this landmark study, the summary statement concluded:

> "…It would appear to be a reasonable option for the clinician to consider the use of Sygen® within the context of clinical studies in selected patients with SCI, particularly those with motor-incomplete injuries" [71].

The 2013 Hadley and Walters cervical spine injury management summary article recommends against the use of GM-1 [56].

**Thyrotrophin Releasing Hormone (TRH)**

TRH has been shown in numerous pre-clinical studies to improve long-term behavioural recovery in animal models of acute SCI [72]. Proposed mechanisms include antagonization of the effects of endogenous opioids, peptidoleukotrienes, platelet-activating factor and excitotoxins. The safety and efficacy of TRH in patients with acute SCI was analysed by performing a double blind, randomised control trial of TRH *vs.* placebo [73]. Only 20 patients were included in the trial 9 with complete and 11 with incomplete SCI. There were no differences in the complete injuries but significantly improved motor and sensory scores in the incomplete group treated with TRH. A larger trial was suggested but has not been performed.

## What Have We Learnt?

The case studies of MPSS, GM-1 and TRH highlight some of the enormous difficulties of converting solid pre-clinical data into absolute benefit in human trials. The challenge often centres on patient selection and outcome measurements. Are all "complete" (*i.e.,* ASIA grade A) injuries the same? Should cervical and thoracic injuries be separated or grouped in analysis? Does the specific level within the cervical or thoracic spinal cord influence prognosis? What about mechanism of injury and energy transfer at the time of injury? Should we be concerned with changes in ASIA grade or is functional outcome, *e.g.,* ability to ambulate at 1 year with a walker, a more important measure of success? Can secondary prevention and reversal of acute SCI really be achieved or should we be focusing on treatment of established chronic SCI, which represents a far larger cohort?

It appears that the studies may have generated more debate and questions than answers and guidelines, at an enormous monetary cost.

## ION CHANNEL BLOCKADE IN SPINAL CORD INJURY

The use of ion channel blockers in acute SCI aims to inhibit excitotoxic neuron death or enhancing neuromuscular stimulus transmission. Various ion channel targets have been tested including potassium, calcium, sodium and glutamate.

### Fampiridine and Potassium Channels

In the presence of an SCI, many neurons traversing the injury site do survive but their function is impaired, contributing to the post-traumatic dysfunction. Demyelination likely plays a major role, as does the altered activity of ion channels [74]. Outward potassium current from voltage-clamped myelinated axons is minimal until the myelin is disrupted, unveiling the potassium channels beneath and effecting cell signal propagation [75]. Potassium channel blockade should therefore improve conduction in demyelinated axons.

Fampiridine (4-aminopyridine) is a blocker of rapidly activating voltage-gated potassium channels. No trial has assessed this drug's efficacy in acute SCI. The first reported trial in chronic SCI was from McMaster University, Canada in 1993.

Five of the six patients with incomplete SCI noticed significant temporary neurological improvement [76]. A similar pre-clinical trial showed enhanced volitional EMG interference patterns in incomplete chronic SCI patients [77]. In chronic quadriplegic patients, 4-aminopyridine has also been shown to improve pulmonary function [78] and reduces low frequency heart rate variability [79]. A double blind, placebo controlled crossover trial using Fampiridine in a series of ambulatory (*i.e.,* incomplete) SCI patients has been performed [80]. In this small study (n = 15) of chronic SCI patients there was no objective evidence of improvement of muscle force, gait dynamics or electromyographic activation patterns despite patients subjective impression of benefit in the treatment group.

A phase 2, double-blind, randomised controlled trial of Fampiridine in chronic incomplete SCI enrolled 91 patients from 11 academic centres in North America [81]. Subjective global patient-reported symptoms and spasticity measures favoured the low-dose treatment group (25mg BID) and more drug-related complications were seen in the high-dose treatment group (40mg BID).

## Gacyclidine and the NMDA Receptor

Gacyclidine (GK-11, Beaufour-Ipsen) is a phencyclidine derivative. It is an N-Methyl-D-Aspartate (NMDA) receptor antagonist with potential neuroprotective benefits [82, 83] and also has a low affinity for muscarinic and opiate receptors [84]. Radio-labelled NMDA receptors and gacyclidine co-locate in a uniform low distribution throughout the rat spinal cord [85].

In an experimental rat ischemic SCI model, intravenous gacyclidine therapy was able to reduce the secondary injury, reduce the lesion size and improve hind limb motor recovery [86]. Similar benefits were shown in a contusive model of rodent SCI [87].

Phase 2 trials for gacyclidine in acute SCI gave disappointing results and the development of the drug for this indication has been discontinued [88]. A phase 2 prospective multi-centre double-blind trial assessed the safety and efficacy of intravenous gacyclidine in acute traumatic brain injury and identified a long-term benefit in functional recovery [89].

## Nimodipine and Calcium Channels

Nimodipine is a dihydropyridine L-type calcium channel blocker. In a rat thoracic model of acute SCI, nimodipine infusion treatment reduced the changes seen in spinal cord evoked potential (SCEP) latency and amplitude. Further, the treatment group had less axonal damage, loss of myelin based protein and ultrastructural myelin vesiculation [90].

Petitjean and Pointillart showed that continuous intravenous nimodipine improved spinal cord blood flow and reduced the size of the SCI lesion in a baboon model of lumbar SCI [91]. Eight years later the same authors then completed a phase 2 human clinical trial of nimodipine *vs.* MPSS, both or no treatment in the acute phase of SCI. Despite enrolling 100 patients, no benefit of nimodipine was detected at 1 year [64]. The heterogeneity of the patient population (complete and incomplete lesions; cervical and thoracic injuries) may have indeed masked any small benefit and reiterates the importance of cautious patient selection in SCI trials.

## HP184 and Sodium/Potassium Channels

HP184 is a sodium/potassium channel blocker that may improve conductance of demyelinated axons. A phase 2 randomised controlled trial sponsored by Sanofi-Aventis assessed the efficacy and safety of a 24 week course of oral HP184 in adults with chronic SCI [92]. 262 patients were enrolled and the trial was completed in 2005 without a published report.

Drug trials have the potential to put the patients at risk as well as benefit them. Publishing results of clinical trials – good or bad – is essential to share knowledge of success and prevent costly errors being reproduced. There is an ethical, but not legal, responsibility to publish such findings.

## Riluzole and Sodium Channels

Riluzole is a benzothiazole anticonvulsant that works *via* sodium channel blockade and inhibits neuronal excitotoxicity [93]. Persistent activation of voltage-sensitive sodium channels is associated with cellular toxicity and may contribute to secondary injury in acute SCI. A rat thoracic model of acute SCI treated with riluzole showed significant improvement in motor function and behaviour with the return to pre-injury levels of somatosensory evoked potentials (SEP). The SEP in the control group did not improve [94]. In a similar separate study, riluzole was found to improve mitochondrial function and to enhance glutamate and glucose uptake [95]. Schwartz and Fehlings re-confirmed the positive effect of riluzole in a rat clip-compression model of SCI [96].

Lang-Lazdunski utilised an aortic cross-clamp rabbit model of acute ischemic SCI and showed that riluzole prevented paraplegia in 9 of 10 subjects [97]. The same model was also used to compare riluzole to magnesium sulfate and again, riluzole-treated rats had a favourable neurological outcome [98]. Riluzole has also been shown to improve neuromotor function, reduce apoptotic neuronal cell death and DNA fragmentation in a rat ischemic SCI model [99].

A phase 1 prospective observational trial by the North American Clinical Trials Network has completed enrolment, with the aim of testing the drug's safety and efficacy in acute traumatic SCI [100]. A large phase 2 double-blind randomised controlled trial is due to begin in January 2013 [101].

## TARGETED THERAPEUTICS IN SPINAL CORD INJURY

Ion channel blockade has to date shown little benefit in human trials of SCI. A great deal of pre-clinical research has since focused on novel therapeutic targets with increased specificity to the spinal neurones and supporting cells. Such drugs include minocycline, cethrin® and SUN13837.

### Minocycline

Minocycline is a broad-spectrum bacteriostatic tertracycline antibiotic that works *via* metalloproteinase inhibition. It is lipophilic, allowing it to penetrate the BSCB. Minocycline has very convincing pre-clinical data. It has been shown to reduce oligodendrocyte death, attenuate axonal dieback, reduce lesion size and improve functional outcome in rat models of acute low cervical SCI [102]. Similar results have been found in a rat ischemic SCI model [103] and a mouse extradural thoracic spinal cord compression model [104]. However, such positive results have been difficult to replicate in a cervical SCI rodent model [105].

Furthermore, a National Institute of Health funded independent replication of the pro-minocycline results published by Lee *et al.* 2003 [106] failed to identify any benefit of the test drug [107].

A phase 2 study from Calgary, Canada is registered with U.S. National Institute of Health clinical trials website [108]. This study compares intravenous minocycline against normal saline placebo, with or without protocol-driven fluid and inotrope augmentation and spinal cord perfusion pressure monitoring. No clinical results have been published but the results of CSF analysis from this study did not show any reduction in cellular prion protein when patients were treated with minocycline [109]. The role of cellular prion protein in spinal cord injury is poorly understood at this time.

## Cethrin®, the Rho Pathway and a Bioengineered Delivery Strategy

Cethrin® (BA-210, Alseres Inc) is a synthetic C3 transferase subunit of botulinum toxin [110], bioengineered to efficiently cross the BSCB [111]. Cethrin® is loaded into fibrin glue and applied directly applied by the surgeon to the subdural injury site. The fibrin solidifies and the cethrin® is slowly released, providing a sustained local delivery.

There are a number of growth inhibitory signals that play a key role in preventing central nervous system re-myelination and axonal sprouting. Among these is a small GTPase, Rho. Rho is activated by axonal contact with growth inhibitory molecules, such as myelin and glial scars [112, 113]. Many pre-clinical studies have identified Rho as a key downstream intracellular enzyme modulating growth inhibition and is activated for at least 7 days after SCI [114]. A key receptor signalling mechanism to Rho is the Nogo receptor (NgR), which is GPI-linked [110]. Activated Rho has a number of downstream targets including Rho Kinase (ROCK) which induces growth cone collapse and neurone retraction [110]. Y-27632 is a ROCK inhibitor that has been shown to promote growth of neurones *in vitro*, even on inhibitory substances [115]. Many rodent models of SCI that have inhibited the Rho pathway have had positive outcomes for spinal cord regeneration, functional recovery and attenuation of allodynia [112, 115-120].

A phase 1/2a clinical study of locally applied Cethrin® was registered with the National Institute of Health in January 2008. A total of 48 patients with acute thoracic or cervical SCI were recruited for this dose-finding study. Pharmacokinetic analysis identified low levels of systemic exposure to the drug and no serious adverse events were attributed to the drug. A positive motor-recovery dose-response relationship was also identified and further, larger, multi-centred clinical trials are planned [121].

Cethrin® is an excellent example of a drug trial based upon solid pre-clinical data that utilised a novel target and a bioengineered local delivery strategy.

## SUN13837 and the FGF Pathway

In the uninjured rodent spinal cord, acidic Fibroblast Growth Factor (aFGF, FGF1) localises to the cytoplasm of ventral motor neurons and dorsal column sensory fibres. Conversely, basic Fibroblast Growth Factor (bFGF, FGF2) immunoreactivity is found in the astrocyte nuclei of a sparse number of neurons in the intermediate gray matter. Two days following injury, the rodent model shows increased aFGF in the ventral motor neurons and a new presence in the intermediate grey matter. Astrocytes are activated by aFGF, which is released from cells during oxidative stress [122]. Interestingly, immunohistochemical identification of bFGF does not show an increase until 5 days post injury [123] but its mRNA expression appears to increase within 6 hours [124]. A rat thoracic model of acute SCI showed that locally applied bFGF improved hind limb motor recovery in combination with methylprednisolone [125]. In a different rat thoracic injury model, Lee *et al.* used a continuous intralesional catheter infusing bFGF and showed a reduction in the size of the lesion and zone of injury [126].

SUN13837 (Asubio Pharmaceuticals Inc.) is a novel therapeutic, which has begun recruitment for a phase 2b trial in North America. The test drug has similar activities to bFGF, affecting the fibroblast growth factor receptor pathway. It is believed to benefit acute SCI *via* neuroprotection from glutamate excitotoxicity. Unlike bFGF, SUN13837 does not stimulate cell proliferation. No pre-clinical data is available on PubMed under the search term "SUN13837" to confirm the company's claim that the study drug has shown benefit in animal models of acute SCI [127]. Patients with acute cervical (C4-C7) SCI with ASIA A impairment are to be randomised in this double-blind trial to receive SUN13837 or placebo within 12 hours of injury for 28 days [128].

A nine patient phase 1 clinical trial of directly applied aFGF was performed in Taiwan and published in 2008 [129]. The patients had chronic cervical SCI (greater than 5 months) and were reassessed monthly post surgery until 6 months. Patients showed improvements in ASIA score, pinprick and light touch scores during the follow up period. No adverse effects were reported but the results have no control group to compare the natural history of recovery so early in the "chronic" phase of SCI.

## ANTIBODY TARGETING OF MYELIN-ASSOCIATED INHIBITION

Systemically or locally administered intravenous antibody therapies are an example of the transition to targeted therapeutics. Antibodies including anti-Nogo-A (Novartis) and anti-MAG (GlaxoSmithKline) aim to reduce myelin-associated inhibition of neuronal growth cone sprouting.

### Anti Nogo-A

Nogo is a myelin-associated axonal growth cone inhibitor found in oligodendrocytes, but not Schwann cells and associates primarily with the endoplasmic reticulum [130]. This has been confirmed with mRNA studies of rat spinal cord tissue, which also showed reduced expression in the lesion epicentre in rat weight-drop model of SCI [131]. Experimental rat acute SCI models have shown improvement in functional recovery with the use of IN-1, a monoclonal antibody raised against Nogo-A [132, 133]. Interestingly, anti Nogo-A antibodies have a growth modulating effect on the intact adult rat cerebellum where Purkinje cells up regulated immediate early genes and grew nodal sprouts along their axons [134, 135].

In 2006, Novartis registered a phase 1 clinical trial of intrathecal administration of ATI355, an anti Nogo-A antibody [136]. The study enrolled thoracic and cervical acute SCI patients of ASIA grade A-C, with treatment to begin within 14 days of injury. Over 50 patients were enrolled, with no side effects described and a phase 2 trial was being planned [137]. This study was completed in November 2011 but no formal results have ever been published. Unfortunately, it appears that this represents another failure to translate good pre-clinical data into patient benefit.

## Anti MAG

Myelin-Associated Glycoproteins (MAG) are myelin-associated neuronal inhibitory substances that cause growth cone collapse [138]. MAG and Nogo share common receptors (NgR1, PirB) and MAG activates the Rho-kinase pathway and down regulates microtubule assembly – an observation that is consistent in SCI [139]. Further, a series of genetically modified knock-out mice showed that NogoA, MAG and oligodendrocyte myelin glycoprotein (OMgp) work in synergy and improve locomotor recovery in a triple knock-out mouse model of SCI [140]. Unfortunately, a very similar study did not confirm the same results [141]. This may, in part, be due to MAG's ability to promote resistance to axonal injury and degeneration in cell culture and *in vivo* [142].

In 2008, GlaxoSmithKline registered the first phase 1 human trial of anti MAG (GSK249320) [143]. This was a single dose, placebo controlled trial in normal healthy subjects. The study has been completed but no published results have documented any complications. GSK249320 has not been studied in human SCI patients but has been trialled (phase 1) in patients with cerebrovascular accidents (stroke). Again, no data has been published.

## MODULATION OF THE ENDOCRINE AND IMMUNE RESPONSE TO SCI

Cervical SCI causes a total separation of the sympathetic, and partial removal of the parasympathetic nervous systems from the brain. The downstream effects include bradycardia, altered coagulation, blood pressure instability and neurohormonal dysregulation [144]. Steroid hormones act at the level of RNA and protein synthesis to effect neuron survival, maintenance of axonal processes and neurotransmission [145].

Neuro-endocrine systems and the immune response should be considered to have an integrated relationship, with many communication pathways. SCI patients have decreased natural and adaptive immune function and reduced serum levels of cell adhesion molecules – important for wound healing and cellular repair mechanisms. Catecholamines, corticosteroids and endorphins can modulate the intensity of the immune response. The reciprocation from the immune system is the release of neurologically active peptides (*e.g.,* adrenocorticotrophic hormone, growth hormone, thyrotrophin, prolactin, somatostatin) [146]. Campagnolo *et al.* showed that quadriplegic patients had higher levels of dehydroepiandrosterone (DHEA) compared to uninjured controls or paraplegic patients and may contribute to the higher level of lifelong infections in this cohort [147].

## Growth Hormone

Growth Hormone (GH) is secreted by the anterior pituitary under stimulation from GH releasing hormone (GHRH) from the hypothalamus. In turn, GH stimulates the liver to produce insulin-like growth factor 1 (IGF-1). A rat T1 clip-compression model of SCI showed some benefit in hind limb motor function recovery after 3 weeks of growth hormone treatment [148]. Exogenous GH has also been shown to preserve SCEP amplitude [149], reduce cord oedema and improve the BSCB after rodent SCI [150]. Bauman *et al.* analysed 9 pairs of monozygotic twins in which one of the twins was paraplegic. In the SCI twin, the GH response to arginine administration was reduced compared to the uninjured twin [151].

No published human trial has assessed the role of GH in acute SCI. An ongoing Spanish phase 3 clinical trial comparing GH and placebo in incomplete chronic SCI patients is aiming to enrol 76 patients and is due for completion in December 2012 [152].

## Oxandrolone

Oxandrolone is an anabolic steroid derived from dihydrotestosterone. It has been primarily used to maintain muscle mass in HIV patients and those with serious burns. Oxandrolone acts *via* the androgen receptor (AR) and inhibits glucocorticoid signalling through crosstalk between AR and the glucocorticoid receptor [153].

Oxandrolone has been shown to improve functional recovery in rat model of acute SCI with the additional observation of axonal sprouting from the ventral horns distal to the injury site [154].

In humans, oxandrolone has been used to successfully treat non-healing pressure ulcers in a small cohort of SCI patients [155]. No published human trial has assessed the benefit of oxandrolone in acute SCI. A study assessing the use of oxandrolone on lean body mass, strength and respiratory function in chronic SCI patients was completed in 2006 but no data has been published [156].

## Immunoglobulin G

Immunoglobulin G (IgG) is an antibody isotype produced by B cells. As part of the secondary injury in SCI, the cellular infiltrative component plays an important role by secreting reactive oxygen species, matrix-metalloproteinase and pro-inflammatory cytokines. IgG has been shown to reduce the activity of leucocytes and microglia in the treatment of autoimmune disease (*e.g.,* Guillain-Barré Syndrome) and therefore may be a potential tool in the treatment of acute SCI [157]. The mechanisms underlying the immunomodulation of IgG are unclear. IgG has been shown to induce monocyte, lymphocyte and neutrophil apoptosis [158, 159] and has a low affinity for complement factor 5a, a potent neutrophil chemoattractant [160]. In addition, the FcY receptors IIIa and IIb are expressed on the surface of neutrophils, macrophages, mast cells, B-lymphocytes and natural killer cells [161].

A rat model of acute SCI identified a significant functional improvement, reduced myeloperoxidase activity (a marker of neutrophils and inflammation) and preserved ultrastructure on electron microscopy when treated with IgG [162].

The role of IgG and immunomodulation in the management of SCI is still in its infancy and lacks replicated results in animal models of SCI. Further pre-clinical research is required in advance of early phase human trials.

## DIRECTLY APPLIED BIOMATERIALS AND LOCAL ENVIRONMENT MODIFICATION

The ability to directly apply a novel bioengineered material to the acutely injured spinal cord would allow the delivery of high local doses of test drug and minimal systemic redistribution. The use of fibrin glue as a drug delivery device has already been detailed in this chapter (see section "Cethrin"). There are multiple other biomaterials that are undergoing pre-clinical development and testing.

## Scaffolds, Hydrogels and Self-Assembling Peptides

Following SCI, a cystic lesion forms within the spinal cord itself and represents an area of cell death, glial scar and neuronal retraction. A therapeutic target therefore would be the introduction of a drug, which provides a physical bridge, or scaffold, across this cavity.

Peripheral nerve grafts can provide a regenerative scaffold for central nerve axons [163]. Bioengineered scaffolds are porous polymers that can be loaded with neural precursor cells and can improve motor function in hemisection models of acute SCI [164]. The disadvantage of these types of scaffolds is that the implantation of a solid structure into human spinal tissue is unfeasible and unlikely to ever gain ethics approval for study.

A novel strategic work-around has been the development of various injectable hydrogels and self-assembling peptides. One particular hydrogel of interest is a blend of hyaluronic acid and methylcellulose (HAMC). This substance degrades slowly when injected into the subarachnoid space, allowing controlled release of any drug loaded onto the gel [165, 166]. Furthermore, the components of the gel itself appear to modulate inflammation and improve neurobehavioural outcomes in a rat model of post-traumatic syringomyelia [167].

Self-assembling peptides have been developed to form nanofibers, nanotubes, surface coatings or vesicles. Their design allows them to be used for potentially many purposes, including 3D cell culture and *in vivo* tissue repair [168]. Peptide amphiphile molecules that self-assemble into nanofibers *in vivo* were injected into a mouse model of acute SCI. This reduced astrogliosis and cell death, promoted axon elongation across injury site and improved neurobehavioural outcomes [169].

The glial scar presents a structural barrier to neuronal repair in the chronic SCI patient. Chondroitinase ABC is an enzyme that is able to break down the glial scar and in turn promote neurite outgrowth, collateral sprouting, synaptic plasticity and axonal regeneration [170, 171]. This is associated with improvement in neurobehavioural outcomes in a chronic mouse model of SCI [172]. Chondroitinase ABC appears to also complement the role of neural progenitor cells and growth factors when applied concurrently to a chronically injured rat spinal cord [173].

## THE EMERGENCE OF STEM CELLS IN SCI

### The Pre-Clinical Promise and Hope for the Future

A stem cell is a self-replicating, progenitor cell with the potential to have its lineage driven to a certain cell type by exposure to different environments, cell signals and transcription factors. Our understanding of stem cell isolation, control and fate determination is progressing rapidly and the potential for their use in many areas of clinical medicine is driving this research from multiple avenues. Stem cells in human randomised clinical trials have shown benefit in reducing acute renal transplant rejection [174], improving cardiac function in ischemic cardiac myopathy [175-177], and possibly even slow the progression of femoral head collapse in avascular necrosis of the femoral head [178]. A number of early phase trials utilising stem cells are registered.

It is the hope of SCI researchers and surgeons that autologous stem cell transplantation, either intravenous or directly into the spinal cord, may be the keystone in the bridge from bench to bedside. The rationale is the innate ability of the cells to respond to the complex pathophysiology of SCI. The cells have the potential to:

1. Modulate endogenous repair mechanisms by minimising cell loss *via* immunomodulation (umbilical cord blood cells).

2. Provide neuroprotective trophic support (bone marrow stromal cells).

3. Modify the local environment by providing scaffolding or remyelination (Schwann cells or olfactory ensheathing cells).

4. Direct cell replacement (*e.g.,* neural precursor cells).

Many stem cell types have shown positive effect in animal models of SCI. In particular olfactory ensheathing cells [179, 180], autologous sural nerve preparations [181, 182] and bone marrow mesenchymal stem cells have shown benefit [183, 184]. Neural precursor cells' fate tends towards oligodendrocyte formation *in vivo* and has shown consistent benefits in rodent models of SCI in terms of lesion size, axon regeneration and neurobehavioural outcomes [173, 185]. Proof of principle of NPCs' potential has been shown by Windrem *et al.*, who were able to repopulate the CNS myelin of hypomyelinated *shiverer* mice and in doing so prevented death in this lethal genetic mutation [186].

## Non-Neural Stem Cell Trials

The first trial of stem cell therapy in human SCI patients was published in 2005. Following good pre-clinical data, Knoller *et al.* used *ex vivo* preparations of autologous macrophages and injected them intra-spinaly, just caudal to the SCI lesion within 2 weeks of injury [187]. This phase 1 study had two cases of non-lethal pulmonary embolism and one case of osteomyelitis, neither of which were attributed to the intervention. Sponsored by Proneuron Biotechnologies, the phase 2 ProCord trial was stopped in 2003 following financial constraints [188]. This trial suffered particularly from the "funnel effect" with only 50 patients meeting the trial entry criteria of the 1816 candidates screened.

Bone Marrow Stromal Cell (BMSC) transplantation has been used in registered phase 1 trials monitoring their safe use in acute and chronic SCI [189, 190]. Small case series evaluating the effect of BMSC transplantation on motor scores and nerve conduction have also been published [191-193]. These early results suggest that intra-thecal delivery of BMSCs is safe. Larger phase 2 randomised controlled trials from Korea (35 acute SCI patients) and Russia (18 chronic SCI patients) have suggested improvements in ASIA scores and again confirmed the safety of the treatment [194, 195]. Interestingly, Yoon *et al.* combined BMSC transplantation with granulocyte macrophage-colony stimulating factor (GM-CSF) in their Phase I/II trial [194]. This builds on the *in vitro* and rodent SCI model analysis of GM-CSF which showed neuronal protection of apoptosis and improved neurologic function [196].

China has two currently registered trials of BMSCs in SCI and three evaluating Umbilical Cord Blood cells (UBC). Globally, these are the only registered UCB trials in SCI and with a planned study size of 100 patients, the results are eagerly awaited.

Olfactory ensheathing cells (OEC) are capable of rapid amplification, migration and myelin production. Autologous OEC transplantation has been performed in SCI case studies and randomised controlled trials [197, 198]. The Australian RCT, headed by Mackay-Sim, enrolled 6 chronic SCI patients who were equally divided between control and treatment groups. There were no deleterious effects from the intra-thecal transplantation at one year [198] and three years [199], however no functional or MRI improvements were identified.

Over 400 chronic SCI patients in China have received aborted foetal olfactory bulb OEC transplantation [200]. These interventions have been unregistered, uncontrolled and unpublished. A review of a single surgeon's practice in Beijing identified a lack of clear inclusion and exclusion criteria, myelopathies of various causes and grades and injection sites that did not correlate with the level of the lesion [200]. This is potentially dangerous for patients and for the pursuit of the stem-cell cure for SCI. With China's vast population comes an opportunity for large scale, controlled RCTs in SCI and if their new trend for registering clinical trials continues, the country may play a substantive role in the direction and success of stem cell therapy.

## Oligodendrocyte Progenitor Cells: The Geron Trials

Geron Corporation (Menlo Park, CA) announced FDA approval in 2009 of their proposal to treat 10 patients with a single dose of $2 \times 10^6$ human embryonic stem cell derived oligodendrocyte progenitor cells (OPCs) within 1 to 2 weeks post-injury [201, 202]. OPCs are a type of pluripotent cell-derived neural precursor cell. The trial only included ASIA grade A patients, despite preclinical data suggesting that cell therapy is most beneficial in incomplete SCI [203, 204].

The now infamous trial was stopped in November 2011. Financial losses of over $100 million dollars in 2010 forced Geron Corporation to terminate the study early. Only four patients received the treatment, codenamed GRNOPC1 and no benefit was found [205].

Fortunately, Geron Corporation's failure has not ended the pursuit of the stem cell cure for SCI. Further phase 1 trials have already begun in hope [206].

## REHABILITATION AND CENTRAL PATTERN GENERATORS

The previously discussed neurobiological approaches to SCI aim to bridge the traumatic lesion. Below the lesion, however, lies a series of functional neural networks (central pattern generators, CPG) that when stimulated, may produce significant functional recovery of sensorimotor functions such as walking and urination [207]. For example, it has been shown that vibration activates locomotor CPG and elicits an involuntary step-like behaviour in complete and incomplete SCI patients [208]. Zhang *et al.* combined scar ablation, OEC transplantation with or without tail nerve electrical stimulation in a chronic rat SCI model. In the groups receiving tail nerve electrical stimulation – which can activate the locomotor CPG – a significant improvement in functional recovery was seen compared to the groups not receiving the electrical stimulation [209]. This may represent a new and very exciting strategy for the treatment of chronic SCI.

## CONCLUSIONS

Spinal cord injury is a devastating illness that has afflicted the lives of millions and will continue to do so. However, scientists and physicians will continue in their pursuit to find the cure for SCI. A great deal of change and progress has been witnessed over the past twenty years – from the methylprednisolone trials to biomaterials and stem cells, alone or in combination – and this provides a great hope for the future. Through the use of randomised controlled trials of appropriately selected patients, with safe and efficacious interventions and sensitive outcome measures, we will continue to make great strides forward; as we believe that our patients will one day make great strides too.

## ACKNOWLEDGEMENTS

Declared none

## CONFLICT OF INTEREST

The authors confirm that this chapter content has no conflict of interest.

## REFERENCES

[1]     Fehlings MG, Cadotte DW, Fehlings LN. A series of systematic reviews on the treatment of acute spinal cord injury: a foundation for best medical practice. J Neurotrauma; 28(8): 1329-33.

[2]     Dumont RJ, Okonkwo DO, Verma S, *et al.* Acute spinal cord injury, part I: pathophysiologic mechanisms. Clin Neuropharmacol 2001; 24(5): 254-64.

[3]     Dumont RJ, Verma S, Okonkwo DO, *et al.* Acute spinal cord injury, part II: contemporary pharmacotherapy. Clin Neuropharmacol 2001; 24(5): 265-79.

[4]     Tator CH. Update on the pathophysiology and pathology of acute spinal cord injury. Brain Pathol 1995; 5(4): 407-13.

[5]     Kwon BK, Okon E, Hillyer J, *et al.* A systematic review of non-invasive pharmacologic neuroprotective treatments for acute spinal cord injury. J Neurotrauma 2011; 28(8): 1545-88. PubMed PMID: 20146558. Epub 2010/02/12. eng.

[6]     Kwon BK, Okon EB, Plunet W, *et al.* A systematic review of directly applied biologic therapies for acute spinal cord injury. J Neurotrauma 2011; 28(8): 1589-610.

[7]     Tetzlaff W, Okon EB, Karimi-Abdolrezaee S, *et al.* A systematic review of cellular transplantation therapies for spinal cord injury. J Neurotrauma 2011; 28(8): 1611-82.

[8]     Hughes JT. The Edwin Smith Surgical Papyrus: an analysis of the first case reports of spinal cord injuries. Paraplegia 1988; 26(2): 71-82.

[9]     Wilson JR, Fehlings MG. Management strategies to optimize clinical outcomes after acute traumatic spinal cord injury: integration of medical and surgical approaches. J Neurosurg Sci 2012; 56(1): 1-11.

[10]    Wilczweski P, Grimm D, Gianakis A, Gill B, Sarver W, McNett M. Risk factors associated with pressure ulcer development in critically ill traumatic spinal cord injury patients. J Trauma Nurs 2012; 19(1): 5-10.

[11]    Hoffman J, Salzman C, Garbaccio C, Burns SP, Crane D, Bombardier C. Use of on-demand video to provide patient education on spinal cord injury. J Spinal Cord Med 2011; 34(4): 404-9.

[12]    Weitzner E, Surca S, Wiese S, *et al.* Getting on with life: positive experiences of living with a spinal cord injury. Qual Health Res 2011; 21(11): 1455-68.

[13]    Riggins MS, Kankipati P, Oyster ML, Cooper RA, Boninger ML. The relationship between quality of life and change in mobility 1 year postinjury in individuals with spinal cord injury. Arch Phys Med Rehabil 2011; 92(7): 1027-33.

[14]    Rundquist J, Gassaway J, Bailey J, Lingefelt P, Reyes IA, Thomas J. The SCIRehab project: treatment time spent in SCI rehabilitation. Nursing bedside education and care management time during inpatient spinal cord injury rehabilitation. J Spinal Cord Med 2011; 34(2): 205-15. PubMed

[15]    Verdonck MC, Chard G, Nolan M. Electronic aids to daily living: be able to do what you want. Disabil Rehabil Assist Technol 2011; 6(3): 268-81.

[16]    Knutsdottir S, Thorisdottir H, Sigvaldason K, Jonsson H, Jr., Bjornsson A, Ingvarsson P. Epidemiology of traumatic spinal cord injuries in Iceland from 1975 to 2009. Spinal Cord 2012; 50(2): 123-6.

[17]    Li J, Liu G, Zheng Y, *et al.* The epidemiological survey of acute traumatic spinal cord injury (ATSCI) of 2002 in Beijing municipality. Spinal Cord 2011; 49(7): 777-82.

[18]    Wyndaele JJ. More knowledge of worldwide incidence and epidemiology of spinal cord injury: data from the United States military. Spinal Cord 2011; 49(8): 857.

[19]    Wyndaele M, Wyndaele JJ. Incidence, prevalence and epidemiology of spinal cord injury: what learns a worldwide literature survey? Spinal Cord 2006; 44(9): 523-9.

[20]    Sekhon LH, Fehlings MG. Epidemiology, demographics, and pathophysiology of acute spinal cord injury. Spine (Phila Pa 1976) 2001; 26(24 Suppl): S2-12.

[21]    Devivo MJ. Epidemiology of traumatic spinal cord injury: trends and future implications. Spinal Cord 2012; 50(5): 365-72.

[22]    Lenehan B, Street J, Kwon BK, *et al.* The epidemiology of traumatic spinal cord injury in British Columbia, Canada. Spine (Phila Pa 1976) 2012; 37(4): 321-9.

[23]    Mirza SK, Krengel WF, 3rd, Chapman JR, *et al.* Early *vs.* delayed surgery for acute cervical spinal cord injury. Clin Orthop Relat Res 1999; (359): 104-14.

[24]    Fehlings MG, Vaccaro A, Wilson JR, *et al.* Early *vs.* Delayed Decompression for Traumatic Cervical Spinal Cord Injury: Results of the Surgical Timing in Acute Spinal Cord Injury Study (STASCIS). PLoS ONE; 7(2): e32037.

[25]    Boakye M, Arrigo RT, Hayden Gephart MG, Zygourakis CC, Lad S. Retrospective, propensity score-matched cohort study examining timing of fracture fixation for traumatic thoracolumbar fractures. J Neurotrauma 2012; 29(12): 2220-5.

[26]    Bourassa-Moreau E, Mac-Thiong JM, Ehrmann Feldman D, Thompson C, Parent S. Complications in acute phase hospitalization of traumatic spinal cord injury: does surgical timing matter? The journal of trauma and acute care surgery 2013; 74(3): 849-54.

[27]    Stahel PF, VanderHeiden T, Flierl MA, *et al.* The impact of a standardized "spine damage-control" protocol for unstable thoracic and lumbar spine fractures in severely injured patients: a prospective cohort study. The journal of trauma and acute care surgery 2013; 74(2): 590-6.

[28]    Bransford RJ, Chapman JR, Skelly AC, VanAlstyne EM. What do we currently know about thoracic spinal cord injury recovery and outcomes? A systematic review. J Neurosurg Spine 2012; 17(1 Suppl): 52-64.

[29]    Mac-Thiong JM, Feldman DE, Thompson C, Bourassa-Moreau E, Parent S. Does timing of surgery affect hospitalization costs and length of stay for acute care following a traumatic spinal cord injury? J Neurotrauma 2012; 29(18): 2816-22.

[30]    Wilson JR, Cadotte DW, Fehlings MG. Clinical predictors of neurological outcome, functional status, and survival after traumatic spinal cord injury: a systematic review. J Neurosurg Spine 2012; 17(1 Suppl): 11-26.

[31]    ClinicalTrials.gov. Understanding Clinical Trials: Nastional Institute of Health; National Library of Medicine; US Department of Health and Human Services; 2007 [updated September 20th, 2007]. Available from: http://clinicaltrials.gov/ct2/info/understand.

[32]    Lammertse D, Tuszynski MH, Steeves JD, *et al.* Guidelines for the conduct of clinical trials for spinal cord injury as developed by the ICCP panel: clinical trial design. Spinal Cord 2007; 45(3): 232-42.

[33]    Steeves JD, Lammertse D, Curt A, *et al.* Guidelines for the conduct of clinical trials for spinal cord injury (SCI) as developed by the ICCP panel: clinical trial outcome measures. Spinal Cord 2007; 45(3): 206-21.

[34]    Fawcett JW, Curt A, Steeves JD, *et al.* Guidelines for the conduct of clinical trials for spinal cord injury as developed by the ICCP panel: spontaneous recovery after spinal cord injury and statistical power needed for therapeutic clinical trials. Spinal Cord 2007; 45(3): 190-205. PubMed

[35]    Tuszynski MH, Steeves JD, Fawcett JW, *et al.* Guidelines for the conduct of clinical trials for spinal cord injury as developed by the ICCP Panel: clinical trial inclusion/exclusion criteria and ethics. Spinal Cord 2007; 45(3): 222-31.

[36]    Sim I, Chan AW, Gulmezoglu AM, Evans T, Pang T. Clinical trial registration: transparency is the watchword. Lancet 2006; 367(9523): 1631-3.

[37]   De Angelis CD, Drazen JM, Frizelle FA, *et al.* Is this clinical trial fully registered?--A statement from the International Committee of Medical Journal Editors. N Engl J Med 2005; 352(23): 2436-8.

[38]   Krleza-Jeric K. Clinical trial registration: the differing views of industry, the WHO, and the Ottawa Group. PLoS Med 2005; 2(11): e378.

[39]   Rockhold FW, Krall RL. Clinical trials registration. PLoS Med 2006; 3(3): e157; author reply e67.

[40]   Krall RL, Rockhold F. Clinical trials report card. N Engl J Med 2006; 354(13): 1426-9; author reply -9.

[41]   Huic M, Marusic M, Marusic A. Completeness and changes in registered data and reporting bias of randomized controlled trials in ICMJE journals after trial registration policy. PLoS One 2011; 6(9): e25258.

[42]   Sances A, Jr., Myklebust JB, Maiman DJ, Larson SJ, Cusick JF, Jodat RW. The biomechanics of spinal injuries. Crit Rev Biomed Eng 1984; 11(1): 1-76.

[43]   Tator CH, Fehlings MG. Review of the secondary injury theory of acute spinal cord trauma with emphasis on vascular mechanisms. J Neurosurg 1991; 75(1): 15-26.

[44]   Waxman SG. Demyelination in spinal cord injury. J Neurol Sci 1989; 91(1-2): 1-14.

[45]   Young W. The post-injury responses in trauma and ischemia: secondary injury or protective mechanisms? Cent Nerv Syst Trauma 1987; 4(1): 27-51.

[46]   Janssen L, Hansebout RR. Pathogenesis of spinal cord injury and newer treatments. A review. Spine (Phila Pa 1976) 1989; 14(1): 23-32.

[47]   Martirosyan NL, Feuerstein JS, Theodore N, Cavalcanti DD, Spetzler RF, Preul MC. Blood supply and vascular reactivity of the spinal cord under normal and pathological conditions. J Neurosurg Spine 2011; 15(3): 238-51.

[48]   Popa C, Popa F, Grigorean VT, *et al.* Vascular dysfunctions following spinal cord injury. J Med Life 2010; 3(3): 275-85.

[49]   Bethea JR. Spinal cord injury-induced inflammation: a dual-edged sword. Prog Brain Res 2000; 128: 33-42.

[50]   Bartanusz V, Jezova D, Alajajian B, Digicaylioglu M. The blood-spinal cord barrier: morphology and clinical implications. Ann Neurol 2011; 70(2): 194-206.

[51]   Popovich PG, Horner PJ, Mullin BB, Stokes BT. A quantitative spatial analysis of the blood-spinal cord barrier. I. Permeability changes after experimental spinal contusion injury. Exp Neurol 1996; 142(2): 258-75.

[52]   Horner PJ, Popovich PG, Mullin BB, Stokes BT. A quantitative spatial analysis of the blood-spinal cord barrier. II. Permeability after intraspinal fetal transplantation. Exp Neurol 1996; 142(2): 226-43.

[53]   Maikos JT, Shreiber DI. Immediate damage to the blood-spinal cord barrier due to mechanical trauma. J Neurotrauma 2007; 24(3): 492-507.

[54]   Noble LJ, Wrathall JR. Distribution and time course of protein extravasation in the rat spinal cord after contusive injury. Brain Res 1989; 482(1): 57-66.

[55]   Cohen DM, Patel CB, Ahobila-Vajjula P, *et al.* Blood-spinal cord barrier permeability in experimental spinal cord injury: dynamic contrast-enhanced MRI. NMR Biomed 2009; 22(3): 332-41.

[56]   Hadley MN, Walters BC. Introduction to the Guidelines for the Management of Acute Cervical Spine and Spinal Cord Injuries. Neurosurgery 2013; 72 Suppl 2: 5-16.

[57]    Markandaya M, Stein DM, Menaker J. Acute Treatment Options for Spinal Cord Injury. Curr Treat Options Neurol 2012.

[58]    Tator CH. Review of treatment trials in human spinal cord injury: issues, difficulties, and recommendations. Neurosurgery 2006; 59(5): 957-82; discussion 82-7.

[59]    Bracken MB, Shepard MJ, Collins WF, Jr., *et al.* Methylprednisolone or naloxone treatment after acute spinal cord injury: 1-year follow-up data. Results of the second National Acute Spinal Cord Injury Study. J Neurosurg 1992; 76(1): 23-31.

[60]    Bracken MB, Shepard MJ, Holford TR, *et al.* Methylprednisolone or tirilazad mesylate administration after acute spinal cord injury: 1-year follow up. Results of the third National Acute Spinal Cord Injury randomized controlled trial. J Neurosurg 1998; 89(5): 699-706.

[61]    Coleman WP, Benzel D, Cahill DW, *et al.* A critical appraisal of the reporting of the National Acute Spinal Cord Injury Studies (II and III) of methylprednisolone in acute spinal cord injury. J Spinal Disord 2000; 13(3): 185-99. P

[62]    Geisler FH, Coleman WP, Benzel E, Ducker T, Hurlbert RJ. Spinal cord injury. Lancet 2002; 360(9348): 1883; author reply 4.

[63]    Hurlbert RJ, Hamilton MG. Methylprednisolone for acute spinal cord injury: 5-year practice reversal. Can J Neurol Sci 2008; 35(1): 41-5.

[64]    Pointillart V, Petitjean ME, Wiart L, *et al.* Pharmacological therapy of spinal cord injury during the acute phase. Spinal Cord 2000; 38(2): 71-6.

[65]    Geisler FH, Dorsey FC, Coleman WP. Recovery of motor function after spinal-cord injury--a randomized, placebo-controlled trial with GM-1 ganglioside. N Engl J Med 1991; 324(26): 1829-38.

[66]    Schonhofer PS. GM-1 ganglioside for spinal-cord injury. N Engl J Med 1992; 326(7): 493; author reply 4.

[67]    Landi G, Ciccone A. GM-1 ganglioside for spinal-cord injury. N Engl J Med 1992; 326(7): 493; author reply 4.

[68]    Geisler FH, Coleman WP, Grieco G, Poonian D, Sygen Study G. The Sygen multicenter acute spinal cord injury study. Spine (Phila Pa 1976) 2001; 26(24 Suppl): S87-98.

[69]    Geisler FH, Coleman WP, Grieco G, Poonian D, Sygen Study G. Recruitment and early treatment in a multicenter study of acute spinal cord injury. Spine (Phila Pa 1976) 2001; 26(24 Suppl): S58-67.

[70]    Geisler FH, Coleman WP, Grieco G, Poonian D, Sygen Study G. Measurements and recovery patterns in a multicenter study of acute spinal cord injury. Spine (Phila Pa 1976) 2001; 26(24 Suppl): S68-86.

[71]    Fehlings MG, Bracken MB. Summary statement: the Sygen(GM-1 ganglioside) clinical trial in acute spinal cord injury. Spine (Phila Pa 1976) 2001; 26(24 Suppl): S99-100.

[72]    Faden AI, Jacobs TP, Smith MT. Thyrotropin-releasing hormone in experimental spinal injury: dose response and late treatment. Neurology 1984; 34(10): 1280-4.

[73]    Pitts LH, Ross A, Chase GA, Faden AI. Treatment with thyrotropin-releasing hormone (TRH) in patients with traumatic spinal cord injuries. J Neurotrauma 1995; 12(3): 235-43.

[74]    Nashmi R, Fehlings MG. Mechanisms of axonal dysfunction after spinal cord injury: with an emphasis on the role of voltage-gated potassium channels. Brain Res Brain Res Rev 2001; 38(1-2): 165-91.

[75]    Chiu SY, Ritchie JM. Potassium channels in nodal and internodal axonal membrane of mammalian myelinated fibres. Nature 1980; 284(5752): 170-1.

[76] Hansebout RR, Blight AR, Fawcett S, Reddy K. 4-Aminopyridine in chronic spinal cord injury: a controlled, double-blind, crossover study in eight patients. J Neurotrauma 1993; 10(1): 1-18.

[77] Hayes KC, Blight AR, Potter PJ, *et al.* Preclinical trial of 4-aminopyridine in patients with chronic spinal cord injury. Paraplegia 1993; 31(4): 216-24.

[78] Segal JL, Brunnemann SR. 4-Aminopyridine improves pulmonary function in quadriplegic humans with longstanding spinal cord injury. Pharmacotherapy 1997; 17(3): 415-23.

[79] Segal JL, Warner AL, Brunnemann SR, Bunten DC. 4-aminopyridine influences heart rate variability in long-standing spinal cord injury. Am J Ther 2002; 9(1): 29-33.

[80] DeForge D, Nymark J, Lemaire E, *et al.* Effect of 4-aminopyridine on gait in ambulatory spinal cord injuries: a double-blind, placebo-controlled, crossover trial. Spinal Cord 2004; 42(12): 674-85.

[81] Cardenas DD, Ditunno J, Graziani V, *et al.* Phase 2 trial of sustained-release fampridine in chronic spinal cord injury. Spinal Cord 2007; 45(2): 158-68.

[82] Drian MJ, Kamenka JM, Privat A. *In vitro* neuroprotection against glutamate toxicity provided by novel non-competitive N-methyl-D-aspartate antagonists. J Neurosci Res 1999; 57(6): 927-34.

[83] Drian MJ, Kamenka JM, Pirat JL, Privat A. Non-competitive antagonists of N-methyl-D-aspartate prevent spontaneous neuronal death in primary cultures of embryonic rat cortex. J Neurosci Res 1991; 29(1): 133-8.

[84] Vincent JP, Cavey D, Kamenka JM, Geneste P, Lazdunski M. Interaction of phencyclidines with the muscarinic and opiate receptors in the central nervous system. Brain Res 1978; 152(1): 176-82.

[85] Subramaniam S, McGonigle P. Quantitative autoradiographic characterization of the binding of (+)-5-methyl-10,11-dihydro-5H-dibenzo[a,d]cyclohepten-5, 10-imine ([3H]MK-801) in rat brain: regional effects of polyamines. J Pharmacol Exp Ther 1991; 256(2): 811-9.

[86] Pencalet P, Ohanna F, Poulat P, Kamenka JM, Privat A. Thienylphencyclidine protection for the spinal cord of adult rats against extension of lesions secondary to a photochemical injury. J Neurosurg 1993; 78(4): 603-9.

[87] Gaviria M, Privat A, d'Arbigny P, Kamenka J, Haton H, Ohanna F. Neuroprotective effects of a novel NMDA antagonist, Gacyclidine, after experimental contusive spinal cord injury in adult rats. Brain Res 2000; 874(2): 200-9.

[88] Mitha AP, Maynard KI. Gacyclidine (Beaufour-Ipsen). Curr Opin Investig Drugs 2001; 2(6): 814-9.

[89] Lepeintre JF, D'Arbigny P, Mathe JF, *et al.* Neuroprotective effect of gacyclidine. A multicenter double-blind pilot trial in patients with acute traumatic brain injury. Neurochirurgie 2004; 50(2-3 Pt 1): 83-95.

[90] Winkler T, Sharma HS, Stalberg E, Badgaiyan RD, Gordh T, Westman J. An L-type calcium channel blocker, nimodipine influences trauma induced spinal cord conduction and axonal injury in the rat. Acta Neurochir Suppl 2003; 86: 425-32.

[91] Petitjean ME, Pointillart V. [Effects of continuous administration of nimodipine during the acute phase of spinal cord injury in baboons]. Ann Fr Anesth Reanim 1992; 11(6): 652-6.

[92] ClinicalTrials.gov. HP184 in Chronic Spinal Cord Injury: U.S. National Library of Medicine; 2004 [updated August 20, 2088]. Available from: http://clinicaltrials.gov/ct2/show/NCT00093275.

[93]   Mazzone GL, Nistri A. Delayed neuroprotection by riluzole against excitotoxic damage evoked by kainate on rat organotypic spinal cord cultures. Neuroscience 2011; 190: 318-27.

[94]   Stutzmann JM, Pratt J, Boraud T, Gross C. The effect of riluzole on post-traumatic spinal cord injury in the rat. Neuroreport 1996; 7(2): 387-92.

[95]   Mu X, Azbill RD, Springer JE. Riluzole improves measures of oxidative stress following traumatic spinal cord injury. Brain Res 2000; 870(1-2): 66-72.

[96]   Schwartz G, Fehlings MG. Evaluation of the neuroprotective effects of sodium channel blockers after spinal cord injury: improved behavioral and neuroanatomical recovery with riluzole. J Neurosurg 2001; 94(2 Suppl): 245-56.

[97]   Lang-Lazdunski L, Heurteaux C, Vaillant N, Widmann C, Lazdunski M. Riluzole prevents ischemic spinal cord injury caused by aortic crossclamping. J Thorac Cardiovasc Surg 1999; 117(5): 881-9.

[98]   Lang-Lazdunski L, Heurteaux C, Dupont H, Widmann C, Lazdunski M. Prevention of ischemic spinal cord injury: comparative effects of magnesium sulfate and riluzole. J Vasc Surg 2000; 32(1): 179-89.

[99]   Lang-Lazdunski L, Heurteaux C, Mignon A, *et al.* Ischemic spinal cord injury induced by aortic cross-clamping: prevention by riluzole. Eur J Cardiothorac Surg 2000; 18(2): 174-81.

[100]  ClinicalTrials.gov. Safety of Riluzole in Patients with Acute Spinal Cord Injury: U.S. National Library of Medicine; 2009 [updated June 28, 2011]. Available from: http://clinicaltrials.gov/ct2/show/NCT00876889.

[101]  ClinicalTrials.gov. Riluzole in Spinal Cord Injury Study (RISCIS): U.S. National Library of Medicine; 2012 [updated May 11, 2012]. Available from: http://clinicaltrials.gov/ct2/show/NCT01597518.

[102]  Stirling DP, Khodarahmi K, Liu J, *et al.* Minocycline treatment reduces delayed oligodendrocyte death, attenuates axonal dieback, and improves functional outcome after spinal cord injury. J Neurosci 2004; 24(9): 2182-90.

[103]  Takeda M, Kawaguchi M, Kumatoriya T, *et al.* Effects of minocycline on hind-limb motor function and gray and white matter injury after spinal cord ischemia in rats. Spine (Phila Pa 1976) 2011; 36(23): 1919-24.

[104]  Wells JE, Hurlbert RJ, Fehlings MG, Yong VW. Neuroprotection by minocycline facilitates significant recovery from spinal cord injury in mice. Brain 2003; 126(Pt 7): 1628-37.

[105]  Lee JH, Tigchelaar S, Liu J, *et al.* Lack of neuroprotective effects of simvastatin and minocycline in a model of cervical spinal cord injury. Exp Neurol 2010; 225(1): 219-30.

[106]  Lee SM, Yune TY, Kim SJ, *et al.* Minocycline reduces cell death and improves functional recovery after traumatic spinal cord injury in the rat. J Neurotrauma 2003; 20(10): 1017-27.

[107]  Pinzon A, Marcillo A, Quintana A, *et al.* A re-assessment of minocycline as a neuroprotective agent in a rat spinal cord contusion model. Brain Res 2008; 1243: 146-51.

[108]  ClinicalTrials.gov. Minocycline and Perfusion Pressure Augmentation in Acute Spinal Cord Injury 2007 [updated June 4, 2008]. Available from: http://clinicaltrials.gov/ct2/show/NCT00559494.

[109]  Carnini A, Casha S, Yong VW, Hurlbert RJ, Braun JE. Reduction of PrP(C) in human cerebrospinal fluid after spinal cord injury. Prion 2010; 4(2): 80-6.

[110]  McKerracher L, Higuchi H. Targeting Rho to stimulate repair after spinal cord injury. J Neurotrauma 2006; 23(3-4): 309-17.

[111] Lord-Fontaine S, Yang F, Diep Q, *et al.* Local inhibition of Rho signaling by cell-permeable recombinant protein BA-210 prevents secondary damage and promotes functional recovery following acute spinal cord injury. J Neurotrauma 2008; 25(11): 1309-22.

[112] Dubreuil CI, Winton MJ, McKerracher L. Rho activation patterns after spinal cord injury and the role of activated Rho in apoptosis in the central nervous system. J Cell Biol 2003; 162(2): 233-43.

[113] Lehmann M, Fournier A, Selles-Navarro I, *et al.* Inactivation of Rho signaling pathway promotes CNS axon regeneration. J Neurosci 1999; 19(17): 7537-47.

[114] McKerracher L, David S. Easing the brakes on spinal cord repair. Nat Med 2004; 10(10): 1052-3.

[115] Dergham P, Ellezam B, Essagian C, Avedissian H, Lubell WD, McKerracher L. Rho signaling pathway targeted to promote spinal cord repair. J Neurosci 2002; 22(15): 6570-7.

[116] Hara M, Takayasu M, Watanabe K, *et al.* Protein kinase inhibition by fasudil hydrochloride promotes neurological recovery after spinal cord injury in rats. J Neurosurg 2000; 93(1 Suppl): 94-101.

[117] Ramer LM, Borisoff JF, Ramer MS. Rho-kinase inhibition enhances axonal plasticity and attenuates cold hyperalgesia after dorsal rhizotomy. J Neurosci 2004; 24(48): 10796-805.

[118] Fournier AE, Takizawa BT, Strittmatter SM. Rho kinase inhibition enhances axonal regeneration in the injured CNS. J Neurosci 2003; 23(4): 1416-23.

[119] Borisoff JF, Chan CC, Hiebert GW, *et al.* Suppression of Rho-kinase activity promotes axonal growth on inhibitory CNS substrates. Mol Cell Neurosci 2003; 22(3): 405-16. P

[120] Sung JK, Miao L, Calvert JW, Huang L, Louis Harkey H, Zhang JH. A possible role of RhoA/Rho-kinase in experimental spinal cord injury in rat. Brain Res 2003; 959(1): 29-38.

[121] Fehlings MG, Theodore N, Harrop J, *et al.* A phase I/IIa clinical trial of a recombinant Rho protein antagonist in acute spinal cord injury. J Neurotrauma 2011; 28(5): 787-96.

[122] Cassina P, Pehar M, Vargas MR, *et al.* Astrocyte activation by fibroblast growth factor-1 and motor neuron apoptosis: implications for amyotrophic lateral sclerosis. J Neurochem 2005; 93(1): 38-46.

[123] Koshinaga M, Sanon HR, Whittemore SR. Altered acidic and basic fibroblast growth factor expression following spinal cord injury. Exp Neurol 1993; 120(1): 32-48.

[124] Follesa P, Wrathall JR, Mocchetti I. Increased basic fibroblast growth factor mRNA following contusive spinal cord injury. Brain Res Mol Brain Res 1994; 22(1-4): 1-8.

[125] Baffour R, Achanta K, Kaufman J, *et al.* Synergistic effect of basic fibroblast growth factor and methylprednisolone on neurological function after experimental spinal cord injury. J Neurosurg 1995; 83(1): 105-10.

[126] Lee TT, Green BA, Dietrich WD, Yezierski RP. Neuroprotective effects of basic fibroblast growth factor following spinal cord contusion injury in the rat. J Neurotrauma 1999; 16(5): 347-56.

[127] Asubio.Pharmaceuticals.Inc. Research Pipeline SUN13837 2011. Available from: http://www.asubio.com/research/research_pipeline.html.

[128] ClinicalTrials.gov. Study to Evaluate the Efficacy, Safety and Pharmacokinetics of SUN13837 Injection in Adult Subjects with Acute Spinal Cord Injury: National Institute of Health; 2011 [updated June 12, 2012]. Available from: http://clinicaltrials.gov/ct2/show/NCT01502631.

[129] Wu JC, Huang WC, Tsai YA, Chen YC, Cheng H. Nerve repair using acidic fibroblast growth factor in human cervical spinal cord injury: a preliminary Phase I clinical study. J Neurosurg Spine 2008; 8(3): 208-14.

[130] GrandPre T, Nakamura F, Vartanian T, Strittmatter SM. Identification of the Nogo inhibitor of axon regeneration as a Reticulon protein. Nature 2000; 403(6768): 439-44.

[131] Josephson A, Widenfalk J, Widmer HW, Olson L, Spenger C. NOGO mRNA expression in adult and fetal human and rat nervous tissue and in weight drop injury. Exp Neurol 2001; 169(2): 319-28.

[132] Merkler D, Metz GA, Raineteau O, Dietz V, Schwab ME, Fouad K. Locomotor recovery in spinal cord-injured rats treated with an antibody neutralizing the myelin-associated neurite growth inhibitor Nogo-A. J Neurosci 2001; 21(10): 3665-73.

[133] Bregman BS, Kunkel-Bagden E, Schnell L, Dai HN, Gao D, Schwab ME. Recovery from spinal cord injury mediated by antibodies to neurite growth inhibitors. Nature 1995; 378(6556): 498-501.

[134] Zagrebelsky M, Buffo A, Skerra A, Schwab ME, Strata P, Rossi F. Retrograde regulation of growth-associated gene expression in adult rat Purkinje cells by myelin-associated neurite growth inhibitory proteins. J Neurosci 1998; 18(19): 7912-29.

[135] Buffo A, Zagrebelsky M, Huber AB, *et al.* Application of neutralizing antibodies against NI-35/250 myelin-associated neurite growth inhibitory proteins to the adult rat cerebellum induces sprouting of uninjured purkinje cell axons. J Neurosci 2000; 20(6): 2275-86.

[136] ClinicalTrials.gov. Acute Safety, Tolerability, Feasibility and Pharmacokinetics of Intrathecal Administered ATI355 in Patients with Acute SCI 2006 [updated November 2, 2011]. Available from: http://clinicaltrials.gov/ct2/show/NCT00406016.

[137] Zorner B, Schwab ME. Anti-Nogo on the go: from animal models to a clinical trial. Ann N Y Acad Sci 2010; 1198 Suppl 1: E22-34.

[138] Li M, Shibata A, Li C, *et al.* Myelin-associated glycoprotein inhibits neurite/axon growth and causes growth cone collapse. J Neurosci Res 1996; 46(4): 404-14.

[139] Mimura F, Yamagishi S, Arimura N, *et al.* Myelin-associated glycoprotein inhibits microtubule assembly by a Rho-kinase-dependent mechanism. J Biol Chem 2006; 281(23): 15970-9.

[140] Cafferty WB, Duffy P, Huebner E, Strittmatter SM. MAG and OMgp synergize with Nogo-A to restrict axonal growth and neurological recovery after spinal cord trauma. J Neurosci 2010; 30(20): 6825-37.

[141] Lee JK, Geoffroy CG, Chan AF, *et al.* Assessing spinal axon regeneration and sprouting in Nogo-, MAG-, and OMgp-deficient mice. Neuron 2010; 66(5): 663-70.

[142] Nguyen T, Mehta NR, Conant K, *et al.* Axonal protective effects of the myelin-associated glycoprotein. J Neurosci 2009; 29(3): 630-7.

[143] ClinicalTrials.gov. Anti-MAG First Administration to Human 2006. Available from: http://clinicaltrials.gov/ct2/show/NCT00622609.

[144] Claus-Walker J, Halstead LS. Metabolic and endocrine changes in spinal cord injury: II (section 1). Consequences of partial decentralization of the autonomic nervous system. Arch Phys Med Rehabil 1982; 63(11): 569-75.

[145] Jones KJ. Steroid hormones and neurotrophism: relationship to nerve injury. Metab Brain Dis 1988; 3(1): 1-18.

[146] Cruse JM, Keith JC, Bryant ML, Jr., Lewis RE, Jr. Immune system-neuroendocrine dysregulation in spinal cord injury. Immunol Res 1996; 15(4): 306-14.

[147]   Campagnolo DI, Bartlett JA, Keller SE. Influence of neurological level on immune function following spinal cord injury: a review. J Spinal Cord Med 2000; 23(2): 121-8.

[148]   Hanci M, Kuday C, Oguzoglu SA. The effects of synthetic growth hormone on spinal cord injury. J Neurosurg Sci 1994; 38(1): 43-9.

[149]   Winkler T, Sharma HS, Stalberg E, Badgaiyan RD, Westman J, Nyberg F. Growth hormone attenuates alterations in spinal cord evoked potentials and cell injury following trauma to the rat spinal cord. An experimental study using topical application of rat growth hormone. Amino Acids 2000; 19(1): 363-71.

[150]   Nyberg F, Sharma HS. Repeated topical application of growth hormone attenuates blood-spinal cord barrier permeability and edema formation following spinal cord injury: an experimental study in the rat using Evans blue, ([125])I-sodium and lanthanum tracers. Amino Acids 2002; 23(1-3): 231-9.

[151]   Bauman WA, Zhang RL, Spungen AM. Provocative stimulation of growth hormone: a monozygotic twin study discordant for spinal cord injury. J Spinal Cord Med 2007; 30(5): 467-72.

[152]   ClinicalTrials.gov. Efficacy and Safety of Growth Hormone Treatment in Spinal Cord Injury (GHSCI) 2011. Available from: http://clinicaltrials.gov/ct2/show/NCT01329757.

[153]   Zhao J, Bauman WA, Huang R, Caplan AJ, Cardozo C. Oxandrolone blocks glucocorticoid signaling in an androgen receptor-dependent manner. Steroids 2004; 69(5): 357-66.

[154]   Zeman RJ, Bauman WA, Wen X, Ouyang N, Etlinger JD, Cardozo CP. Improved functional recovery with oxandrolone after spinal cord injury in rats. Neuroreport 2009; 20(9): 864-8.

[155]   Spungen AM, Koehler KM, Modeste-Duncan R, Rasul M, Cytryn AS, Bauman WA. 9 clinical cases of nonhealing pressure ulcers in patients with spinal cord injury treated with an anabolic agent: a therapeutic trial. Adv Skin Wound Care 2001; 14(3): 139-44.

[156]   ClinicalTrials.gov. The Use of Anabolic Steroids to Improve Function After Spinal Cord Injury 2005. Available from: http://clinicaltrials.gov/ct2/show/NCT00223769.

[157]   Fehlings MG, Nguyen DH. Immunoglobulin G: a potential treatment to attenuate neuroinflammation following spinal cord injury. J Clin Immunol 2010; 30 Suppl 1: S109-12.

[158]   Prasad NK, Papoff G, Zeuner A, *et al.* Therapeutic preparations of normal polyspecific IgG (IVIg) induce apoptosis in human lymphocytes and monocytes: a novel mechanism of action of IVIg involving the Fas apoptotic pathway. J Immunol 1998; 161(7): 3781-90.

[159]   von Gunten S, Schaub A, Vogel M, Stadler BM, Miescher S, Simon HU. Immunologic and functional evidence for anti-Siglec-9 autoantibodies in intravenous immunoglobulin preparations. Blood 2006; 108(13): 4255-9.

[160]   Basta M, Van Goor F, Luccioli S, *et al.* F(ab)'2-mediated neutralization of C3a and C5a anaphylatoxins: a novel effector function of immunoglobulins. Nat Med 2003; 9(4): 431-8.

[161]   Schmidt RE, Gessner JE. Fc receptors and their interaction with complement in autoimmunity. Immunol Lett 2005; 100(1): 56-67.

[162]   Gok B, Sciubba DM, Okutan O, *et al.* Immunomodulation of acute experimental spinal cord injury with human immunoglobulin G. J Clin Neurosci 2009; 16(4): 549-53.

[163]   David S, Aguayo AJ. Axonal elongation into peripheral nervous system "bridges" after central nervous system injury in adult rats. Science 1981; 214(4523): 931-3.

[164]   Teng YD, Lavik EB, Qu X, *et al.* Functional recovery following traumatic spinal cord injury mediated by a unique polymer scaffold seeded with neural stem cells. Proc Natl Acad Sci U S A 2002; 99(5): 3024-9.

[165] Baumann MD, Kang CE, Stanwick JC, *et al.* An injectable drug delivery platform for sustained combination therapy. J Control Release 2009; 138(3): 205-13.

[166] Wang Y, Lapitsky Y, Kang CE, Shoichet MS. Accelerated release of a sparingly soluble drug from an injectable hyaluronan-methylcellulose hydrogel. J Control Release 2009; 140(3): 218-23.

[167] Austin JW, Kang CE, Baumann MD, *et al.* The effects of intrathecal injection of a hyaluronan-based hydrogel on inflammation, scarring and neurobehavioural outcomes in a rat model of severe spinal cord injury associated with arachnoiditis. Biomaterials 2012; 33(18): 4555-64.

[168] Liu J, Zhao X. Design of self-assembling peptides and their biomedical applications. Nanomedicine (Lond) 2011; 6(9): 1621-43.

[169] Tysseling-Mattiace VM, Sahni V, Niece KL, *et al.* Self-assembling nanofibers inhibit glial scar formation and promote axon elongation after spinal cord injury. J Neurosci 2008; 28(14): 3814-23.

[170] Bradbury EJ, Moon LD, Popat RJ, *et al.* Chondroitinase ABC promotes functional recovery after spinal cord injury. Nature 2002; 416(6881): 636-40.

[171] Bradbury EJ, Carter LM. Manipulating the glial scar: chondroitinase ABC as a therapy for spinal cord injury. Brain Res Bull 2011; 84(4-5): 306-16.

[172] Carter LM, McMahon SB, Bradbury EJ. Delayed treatment with chondroitinase ABC reverses chronic atrophy of rubrospinal neurons following spinal cord injury. Exp Neurol 2011; 228(1): 149-56.

[173] Karimi-Abdolrezaee S, Eftekharpour E, Wang J, Schut D, Fehlings MG. Synergistic effects of transplanted adult neural stem/progenitor cells, chondroitinase, and growth factors promote functional repair and plasticity of the chronically injured spinal cord. J Neurosci 2010; 30(5): 1657-76.

[174] Tan J, Wu W, Xu X, *et al.* Induction therapy with autologous mesenchymal stem cells in living-related kidney transplants: a randomized controlled trial. JAMA 2012; 307(11): 1169-77.

[175] Perin EC, Silva GV, Zheng Y, *et al.* Randomized, double-blind pilot study of transendocardial injection of autologous aldehyde dehydrogenase-bright stem cells in patients with ischemic heart failure. Am Heart J 2012; 163(3): 415-21, 21 e1.

[176] Makkar RR, Smith RR, Cheng K, *et al.* Intracoronary cardiosphere-derived cells for heart regeneration after myocardial infarction (CADUCEUS): a prospective, randomised phase 1 trial. Lancet 2012; 379(9819): 895-904.

[177] Bolli R, Chugh AR, D'Amario D, *et al.* Cardiac stem cells in patients with ischaemic cardiomyopathy (SCIPIO): initial results of a randomised phase 1 trial. Lancet 2011; 378(9806): 1847-57.

[178] Zhao D, Cui D, Wang B, *et al.* Treatment of early stage osteonecrosis of the femoral head with autologous implantation of bone marrow-derived and cultured mesenchymal stem cells. Bone 2012; 50(1): 325-30.

[179] Richter MW, Fletcher PA, Liu J, Tetzlaff W, Roskams AJ. Lamina propria and olfactory bulb ensheathing cells exhibit differential integration and migration and promote differential axon sprouting in the lesioned spinal cord. J Neurosci 2005; 25(46): 10700-11.

[180] Steward O, Sharp K, Selvan G, *et al.* A re-assessment of the consequences of delayed transplantation of olfactory lamina propria following complete spinal cord transection in rats. Exp Neurol 2006; 198(2): 483-99.

[181]   Houle JD, Tom VJ, Mayes D, Wagoner G, Phillips N, Silver J. Combining an autologous peripheral nervous system "bridge" and matrix modification by chondroitinase allows robust, functional regeneration beyond a hemisection lesion of the adult rat spinal cord. J Neurosci 2006; 26(28): 7405-15.

[182]   Lavdas AA, Chen J, Papastefanaki F, *et al.* Schwann cells engineered to express the cell adhesion molecule L1 accelerate myelination and motor recovery after spinal cord injury. Exp Neurol 2010; 221(1): 206-16.

[183]   Lu P, Jones LL, Tuszynski MH. Axon regeneration through scars and into sites of chronic spinal cord injury. Exp Neurol 2007; 203(1): 8-21.

[184]   Zurita M, Vaquero J, Bonilla C, *et al.* Functional recovery of chronic paraplegic pigs after autologous transplantation of bone marrow stromal cells. Transplantation 2008; 86(6): 845-53.

[185]   Cummings BJ, Uchida N, Tamaki SJ, *et al.* Human neural stem cells differentiate and promote locomotor recovery in spinal cord-injured mice. Proc Natl Acad Sci U S A 2005; 102(39): 14069-74.

[186]   Windrem MS, Schanz SJ, Guo M, *et al.* Neonatal chimerization with human glial progenitor cells can both remyelinate and rescue the otherwise lethally hypomyelinated shiverer mouse. Cell Stem Cell 2008; 2(6): 553-65.

[187]   Knoller N, Auerbach G, Fulga V, *et al.* Clinical experience using incubated autologous macrophages as a treatment for complete spinal cord injury: phase I study results. J Neurosurg Spine 2005; 3(3): 173-81.

[188]   Jones LA, Lammertse DP, Charlifue SB, *et al.* A phase 2 autologous cellular therapy trial in patients with acute, complete spinal cord injury: pragmatics, recruitment, and demographics. Spinal Cord 2010; 48(11): 798-807.

[189]   Callera F, do Nascimento RX. Delivery of autologous bone marrow precursor cells into the spinal cord *via* lumbar puncture technique in patients with spinal cord injury: a preliminary safety study. Exp Hematol 2006; 34(2): 130-1.

[190]   Sykova E, Homola A, Mazanec R, *et al.* Autologous bone marrow transplantation in patients with subacute and chronic spinal cord injury. Cell Transplant 2006; 15(8-9): 675-87.

[191]   Moviglia GA, Fernandez Vina R, Brizuela JA, *et al.* Combined protocol of cell therapy for chronic spinal cord injury. Report on the electrical and functional recovery of two patients. Cytotherapy 2006; 8(3): 202-9.

[192]   Park HC, Shim YS, Ha Y, *et al.* Treatment of complete spinal cord injury patients by autologous bone marrow cell transplantation and administration of granulocyte-macrophage colony stimulating factor. Tissue Eng 2005; 11(5-6): 913-22.

[193]   Saito F, Nakatani T, Iwase M, *et al.* Spinal cord injury treatment with intrathecal autologous bone marrow stromal cell transplantation: the first clinical trial case report. J Trauma 2008; 64(1): 53-9.

[194]   Yoon SH, Shim YS, Park YH, *et al.* Complete spinal cord injury treatment using autologous bone marrow cell transplantation and bone marrow stimulation with granulocyte macrophage-colony stimulating factor: Phase I/II clinical trial. Stem Cells 2007; 25(8): 2066-73.

[195]   Chernykh ER, Stupak VV, Muradov GM, *et al.* Application of autologous bone marrow stem cells in the therapy of spinal cord injury patients. Bull Exp Biol Med 2007; 143(4): 543-7.

[196] Ha Y, Park HS, Park CW, *et al.* Synthes Award for Resident Research on Spinal Cord and Spinal Column Injury: granulocyte macrophage colony stimulating factor (GM-CSF) prevents apoptosis and improves functional outcome in experimental spinal cord contusion injury. Clin Neurosurg 2005; 52: 341-7.

[197] Lima C, Pratas-Vital J, Escada P, Hasse-Ferreira A, Capucho C, Peduzzi JD. Olfactory mucosa autografts in human spinal cord injury: a pilot clinical study. J Spinal Cord Med 2006; 29(3): 191-203; discussion 4-6.

[198] Feron F, Perry C, Cochrane J, *et al.* Autologous olfactory ensheathing cell transplantation in human spinal cord injury. Brain 2005; 128(Pt 12): 2951-60.

[199] Mackay-Sim A, Feron F, Cochrane J, *et al.* Autologous olfactory ensheathing cell transplantation in human paraplegia: a 3-year clinical trial. Brain 2008; 131(Pt 9): 2376-86.

[200] Dobkin BH, Curt A, Guest J. Cellular transplants in China: observational study from the largest human experiment in chronic spinal cord injury. Neurorehabil Neural Repair 2006; 20(1): 5-13.

[201] Alper J. Geron gets green light for human trial of ES cell-derived product. Nat Biotechnol 2009; 27(3): 213-4.

[202] Couzin J. Biotechnology. Celebration and concern over U.S. trial of embryonic stem cells. Science 2009; 323(5914): 568.

[203] Faulkner J, Keirstead HS. Human embryonic stem cell-derived oligodendrocyte progenitors for the treatment of spinal cord injury. Transpl Immunol 2005; 15(2): 131-42.

[204] Keirstead HS, Nistor G, Bernal G, *et al.* Human embryonic stem cell-derived oligodendrocyte progenitor cell transplants remyelinate and restore locomotion after spinal cord injury. J Neurosci 2005; 25(19): 4694-705.

[205] Kaiser J. Embryonic stem cells. Researchers mull impact of Geron's sudden exit from field. Science 2011; 334(6059): 1043.

[206] Mack GS. ReNeuron and StemCells get green light for neural stem cell trials. Nat Biotechnol 2011; 29(2): 95-7.

[207] Rossignol S, Schwab M, Schwartz M, Fehlings MG. Spinal cord injury: time to move? J Neurosci 2007; 27(44): 11782-92.

[208] Field-Fote E, Ness LL, Ionno M. Vibration elicits involuntary, step-like behavior in individuals with spinal cord injury. Neurorehabil Neural Repair 2012; 26(7): 861-9.

[209] Zhang SX, Huang F, Gates M, Holmberg EG. Tail nerve electrical stimulation combined with scar ablation and neural transplantation promotes locomotor recovery in rats with chronically contused spinal cord. Brain Res 2012; 1456: 22-35.

*Send Orders for Reprints to reprints@benthamscience.net*

# CHAPTER 4

## Peptide Neurotoxins Targeting Voltage-Gated Ion Channels and Their Therapeutic Implications

**Seungkyu Lee[1,2] and Sun W. Hwang[3,*]**

*[1]F.M. Kirby Neurobiology Center, Children's Hospital Boston, Boston, MA 02115, USA; [2]Department of Neurobiology, Harvard Medical School, Boston, MA 02115, USA and [3]Department of Biomedical Sciences, Korea University College of Medicine, Seoul 136-705, Korea*

**Abstract:** Voltage-gated ion channels are transmembrane proteins that selectively permeate $K^+$, $Na^+$, or $Ca^{2+}$ in response to the plasma membrane voltage changes. They play an essential role in neuronal excitability and molecular signaling in central and peripheral nervous systems. Dysfunctions of these ion channels are involved in various neurological disorders such as pain, migraine, schizophrenia, Alzheimer's disease, epilepsy, depression, *etc.* Thus, increasing numbers of academic and medical institutions and pharmaceutical industries have paid attention to these ion channel proteins as therapeutic targets. Consequently, a number of medicines modulating the channel functions have been being developed. Among the voltage-gated ion channel modulators, peptide neurotoxins from venomous animals such as cone snails, spiders, scorpions, and sea anemones are noticeable in that they modify voltage-gated ion channel activities in highly selective and potent manners. The distinctive selectivity and potency are based on their particularly rigid three-dimensional structures with multiple disulfide bonds, which confer strong and specific binding to a channel subtype. Varied inter-cysteine sequences further give an additional specificity. Ziconotide, blocking neuronal N-type voltage-gated $Ca^{2+}$ channels in this mechanism, is the first FDA-approved peptide toxin for neurological diseases. Several other peptidergic neurotoxins are in preclinical or clinical phases. Here, we update knowledge on molecular and functional characteristics of the peptide neurotoxins targeting voltage-gated ion channels in the nervous system. We also discuss their current status of research and developments and their future therapeutic potentials.

**Keywords:** Peptide neurotoxin, voltage-gated $K^+$ channels, voltage-gated $Na^+$ channels, voltage-gated $Ca^{2+}$ channels, Neurological diseases, molecular properties, therapeutic potentials.

*****Address correspondence to Sun W. Hwang:** Korea University Ansan Hospital #3513, Gojan-1-Dong, Danwon-Gu, Ansan-Shi, Gyeonggi-Do 425-707, Korea; Tel: +82-31-412-6710; Fax: +82-31-412-6729; E-mail: sunhwang@korea.ac.kr

**Atta-ur-Rahman (Ed)**
**All rights reserved-© 2013 Bentham Science Publishers**

# INTRODUCTION

Voltage-gated ion channels are integral proteins with multiple membrane spanning domains. The channels selectively permeate $K^+$, $Na^+$, and $Ca^{2+}$ when the membrane voltage is changed. These voltage-gated channels are indispensable for the functions of excitable cells including neurons, skeletal, smooth and cardiac muscles. For neurons, these channels are essential for firing of action potentials and releases of neurotransmitters towards synaptic clefts no matter where those are located, in central or peripheral nervous system (CNS or PNS). Therefore, it is not surprising that abnormal function of voltage-gated ion channels has been related to diverse neurological defects such as chronic pain, epilepsy, depression, and Alzheimer disease. Thus, these classes of channel proteins constitute a major drug target for diseases of the nervous system. For cardiovascular diseases, numbers of channel-targeted drugs have already been contributing as therapeutics to saving patients' lives [1, 2]. In either case for cardiovascular or nervous system, even when the diseases are with other causes than the ion channel itself or with unknown ones, the channel modulators have been able to significantly reverse pathologic symptoms [3, 4]. Therefore, modulation of voltage-gated channels is becoming more attractive as a therapeutic avenue. Such utility is getting more promising as information of molecular nature and roles of the channel in the nervous system are explosively being increased by currently advanced biotechnology.

Constant efforts in genomic research have discovered a large number of subtypes of voltage-gated ion channels. According to a genome-wide search for genes encoding at least one putative pore-forming region using RefSeq (NCBI Reference Sequence) database and hidden Markov models, 143 genes were revealed, encoding primary α subunits of voltage-gated-like ion channels [5, 6]. These days, such heterogeneity would be several times increased when the splice variants or posttranslational modifications of channels with functional relevance are considered [7, 8]. Although this complexity appears to disperse our focus on the roles and validities of the voltage-gated channels, this might also make it more possible to control a disease better by pinpoint modulation of a certain specific voltage-gated channel with less off-target effects than predicted previously, since each subtype probably has a confined location for expression and also a limited pathologic status in regards to its aberrant function. For example, molecular genetic studies of voltage-gated $Ca^{2+}$ channel (Cav) subtypes demonstrated that T-type and N-type Cav channels are therapeutic targets for absence epilepsy and chronic pain, respectively [9, 10].

In order to harness the function of voltage-gated ion channels and develop novel therapeutics targeting them, one should understand the expression profiles and functions of each subtype of voltage-gated ion channels. Currently, two major methods are being used to understand the roles of voltage-gated ion channels. One is genetic engineering, which includes knock-out, knock-in, and RNA interference (RNAi)-mediated knock-down strategies. The advantage of genetic modification is that interesting genes are modified without off-target effect in the observed system. However, those approaches are time-consuming and relatively high-cost ones. In addition, a compensatory mechanism can occur by alternative expression of other subtypes and by heterotetramer formation. The other method to explore the roles of voltage-gated channels is a pharmacological way in which outcomes from functional changes of target channels by the specific pharmacological blockers or activators are observed. This way is straightforward and fast to obtain results and even further, chemical information of the agents could be directly helpful to develop therapeutics. However, off-target effects should be considered, in particular when high doses are used. Therefore, it would be a better idea to complementarily translate results from both approaches for the understanding of ion channels function than to lean only on the either side.

In the context of off-target effects, peptides are now attracting attentions. The difficulty in lowering failure rate of synthetic small molecules in clinical developments is one of the current issues in pharmaceutical industry. Thus, researchers are turning their gaze on different chemical pools. Coincidently, number of novel peptides from plants, animal venoms, and phage displays are now growing owing to endeavors in biology field. The fact that information on the effects of many of those on the human body is still lacking is stimulating the curiosity of biomedical researchers. Moreover, there is accumulating experience for exploring peptide (toxin)-voltage-gated channel interactions in physiology and pharmacology fields.

In biochemistry, peptides are defined as short polymers connecting below 50 amino acids as building blocks by peptide (amide) bonds. Peptides from animal venoms, which are called peptide neurotoxins, mainly interfere with neuronal activity by modulating ion channel activities and subsequent neuronal communications in synapses. It might be good to note that, in peptide neurotoxin research, the criteria of this peptide length is rather ambiguous so that several groups are sometimes called protein neurotoxins (51-100 amino acids) as well as peptide neurotoxins. In this chapter, protein neurotoxins are incorporated among peptide neurotoxins because the pharmacological and structural characteristics of both neurotoxins are not different except for the size and those neurotoxins are produced by same methods such as solid phase peptide synthesis (SPPS) and bacterial expression. These toxins have been isolated by conventional biochemistry, molecular cloning, and assay-based isolation [11]. The peptide neurotoxins have been found mostly in venom apparatuses of a number of organisms such as cone snails, spiders, scorpions, sea anemones, and snakes. Each animal uses the toxin mixture containing diverse peptides for self-defense and prey capture. Not surprisingly, due to the vital role of voltage-gated ion channels in body physiology mentioned above, the peptide toxins are evolutionarily accommodated to targeting these essential transmembrane proteins in nervous systems. The major common characteristic of peptide neurotoxins is a high proportion of cysteine residues in their primary sequences, which are essential for structural stability and fitting on the binding pocket of voltage-gated ion channels [12-14]. Compared to small synthetic chemicals, peptide neurotoxins would be useful bio-materials because the toxins could be endogenously generated in the body *via* genetic engineering as well as extraneously administered through pharmacological ways.

Until now, very low percentage (~1%) of peptide neurotoxins has been studied. Despite such little investigation, the peptide neurotoxin research has given two definite benefits. Those are structural and functional insights for the voltage-gated ion channels, and the toxin's own therapeutic potentials. In particular, the peptide neurotoxins targeting voltage-gated $K^+$ (Kv) channels and voltage-gated $Na^+$ (Nav) channels have given many crucial insights in channel structures even before elucidating crystal structures of Kv channels [15, 16]. ω-conotoxin MVIIA, one of peptide neurotoxins isolated from cone snail *Conus magus,* has been developed to ziconotide (Prialt®, Elan corp.), a clinical medicine for refractory pain. It appears to be one of the best examples for peptide neurotoxin therapeutics because the drug can alleviate refractory pains that the opioids and NSAIDs fail to control. Considering these contributions of peptide neurotoxins to understanding the nature of voltage-gated ion channel behaviors and therapeutic developments, increasing numbers of peptide libraries by current technological advances must have a big potential and open a new window in drug discovery for CNS diseases. As to interactions between peptide neurotoxins and target voltage-gated ion channels, several useful reviews are available, which mostly focus on the animal sources and target channels [12, 17-20]. In this review, we will focus on and describe molecular and functional characteristics of peptide neurotoxins and their target voltage-gated ion channels, current status of the research and development, and its future perspectives.

## VOLTAGE-GATED POTASSIUM CHANNELS (KV CHANNELS)

### Molecular Properties of Kv Channels

Kv channels have been first cloned from *Drosophila* (fruitfly) and shortly after that, human homologs began to be uncovered [21-23]. Since then, 40 human Kv channels have been characterized and grouped into 12 families according to phylogenetic similarity [24]. In excitable cells, the opening of Kv channels causes outward currents (potassium ion efflux), which repolarize or even hyperpolarize the cell membrane, because the cytosolic concentration of potassium ion ($K^+$) is higher than outside the cells. Maintaining the functions of Kv channels in their normal ranges is essential since they are the major component to set the cellular resting membrane potential and at the same time to finely tune the neuronal excitability in the nervous system. Therefore, the functional disruptions of Kv channels by mutations or by particular metabolites may lead CNS to pathologic stages such as epilepsy, multiple sclerosis, spinal cord injury, neuropathic pain, and Alzheimer's disease [25]. Readjustment of Kv channel activities *via* their pharmacological activations or inhibitions is often helpful to reverse the disease states. For example, retigabine, a Kv7 channel activator that may prevent hyperexcitability of neurons, has been approved by US Food and Drug Administration (FDA) and European Medicines Agency in 2011 for the therapy of partial epilepsies [26]. The fact suggests that Kv channels are clinically worth targeting for treatment of CNS diseases and that selective modulators might come more into the limelight.

A functional Kv channel is a homo- or hetero- tetramer, in which each monomer contains 6 transmembrane segments (S1-S6). Those six segments are divided into two functional domains; pore domain and voltage sensing domain. The pore domain consists of S5, P-loop, and S6, and contains a selectivity filter for potassium ion and a gate. The voltage sensor domain, which consists of S1 to S4, senses changes in the membrane voltage and converts the electrical energy into the pore opening. In particular, the S4 contains positively-charged amino acid residues at every third residue over 4 to 8 amino acids so that it has been known that the S4 movement in response to the membrane voltage changes is a crucial step for the opening of channels. Studies on X-ray structures of KvAP from an archaebacterium *Aeropyrum pernix* [27] and of Kv1.2 [28] by Mackinnon and co-workers revealed the facts that the pore region consists of outer vestibule, a selectivity filter, cavity, and S6 gate in the center and that the four voltage sensing domains of each four subunits are located in the outskirts of the channel complex. A most part of voltage sensing domains interact with lipid of the surrounding cell membrane and only a small part of them contacts the pore domains of neighboring subunits but not those of the same subunit. They also suggested that the voltage sensor paddle, a helix-turn-helix motif present in S3b-S4 region, contacts the membrane lipid bilayer and that this interaction is important for channel movements in response to a voltage change. The voltage sensor paddle is located at inner leaflet of the membrane at resting state. Upon depolarization, the paddle moves to outer leaflet, indicating a large movement (~15Å). This voltage sensor paddle hypothesis competed with the conventional helical screw model, in which S4 moves in a small distance vertically throughout the membrane. At this point, a peptide neurotoxin interacting with the voltage sensing domains provided a critical information correcting the paddle hypothesis, details of which will be explained below [29]. Considering the movement of voltage sensor domain and gating of the pore, Kv channels seem to have different conformations such as open, closed, and inactivated states, depending on particular membrane voltages and their durations. These varied conformations may also have different affinities to a same exogenous ligand. In accordance with this, some chemicals are known to show their use-dependent and state-dependent effect.

## Peptide Neurotoxins for Kv Channels

Throughout the history of Kv channel research, toxins with high affinity and selectivity have been very useful tools in studies on the nature of channel properties. For instance, The high affinity toxins for Kv channels have given insights in understanding Kv channel structures even before determining an X-ray structure. Use of toxins selectively acting on Kv subtypes have made it possible to define physiological roles and to predict pathological roles of given subtypes [30]. In addition, the peptide neurotoxin has been used to purify channel proteins and determine subunit composition [31]. Currently, such progresses in understanding Kv channel properties stimulates not only more detailed delineation of the processes of Kv-related diseases, but also conception and practical validation of therapeutic utilities of a number of peptide neurotoxins for Kv channels. The peptide neurotoxins targeting Kv channels can be classified into pore blockers and gating modifiers according to their binding sites, pore domain and voltage sensing domain [32]. The functions and benefits of these two neurotoxin types will be described below.

### Pore Blockers

The pore blocking toxins bind to residues in the outer pore vestibule of Kv channels with 1:1 stoichiometry. They block ion conduction by physically occluding the pore (Table **1**, Fig. **1**). The toxins usually do not affect the gating process of channels. It is possible that the pore blockers may bind to open and closed channels with different affinities, which results from conformation differences between each state. Although this state dependence of the toxin binding has not been studied thoroughly, there are a few clear examples that the pore blockers such as conotoxins PVIIA and RIIIK showed state-dependent inhibition (preferring open channels for binding) [33]. Noxiustoxin, isolated from a scorpion *Centruroides noxius* in 1982, is the first pore blocker. Since this discovery, a large number of Kv channel toxins have been found and chemically and pharmacologically characterized, from scorpions, spiders, cone snails, sea anemones, and snakes [34, 35]. Among those, charybodotoxin, a 37 amino acid toxin from another scorpion *Leiurus quinquestriatus hebraeus*, is the best studied and became one of the classic examples of pore blockers because of its high affinity (1-3 nM) compared to noxiustoxin (450 nM) against Kv channels [36-39]. Interestingly, diverse scorpion toxins have been found and most of them are shown to target Kvs as pore blockers. They have extensive pools dubbed as KTx family. Individual toxins are known to have Kv subtype specificities (Table **2**) [38, 40]. Studies with the scorpion toxins have demonstrated that the association rates of toxins to channels are decreased in presence of tetraethylammonium (TEA), a small molecule pan $K^+$ channel pore blocker, and that the dissociation rates are increased along the elevation of cytosolic monovalent cation concentrations [41, 42]. These results suggested that the toxins interact with the pore region and raised the possibility that the Kv pore region could be identified using the peptide neurotoxins. Indeed, the pore region of *Drosophila Shaker* Kv channel (which corresponds to human Kv1 channels) has been structurally defined by determining the responsiveness of mutant channels to charybdotoxin [36]. Moreover, it has first been established by the agitoxin experiment that the Kv channel subunits assemble as a tetramer to form a functional channel [43]. In addition, several toxins, structural information of which are available, has been molecular calipers so that the pore dimension that had been deduced from toxin data has well matched the crystal structure of a bacterial primitive $K^+$ channel KcsA [44, 45]. Taken together, the pore blockers isolated from scorpion have given a key contribution to providing structural insights for the pore region of Kv channel together with crystallographic approaches.

**Table 1:** Characteristics of the two classes of peptide neurotoxins for Kv channels

|  | **Pore Blockers** | **Gating Modifiers** |
|---|---|---|
| Binding region | Pore domain (S5, P-loop, and S6) | Voltage sensing domain (S1-S4) |
| Binding stoichiometry (toxin:channel tetramer) | 1:1 | 4:1 |
| Peptide neurotoxins | Charybdotoxin, Agitoxin, κ-PVIIA | Hanatoxin, SGTx1, GxTX-1E |
| Electrophysiology | No shift of voltage activation | Rightward shift of voltage activation |
| Structural topology | CSαβ, ICK | ICK |
| Surface profile | Positively residues | Hydrophobic patch |

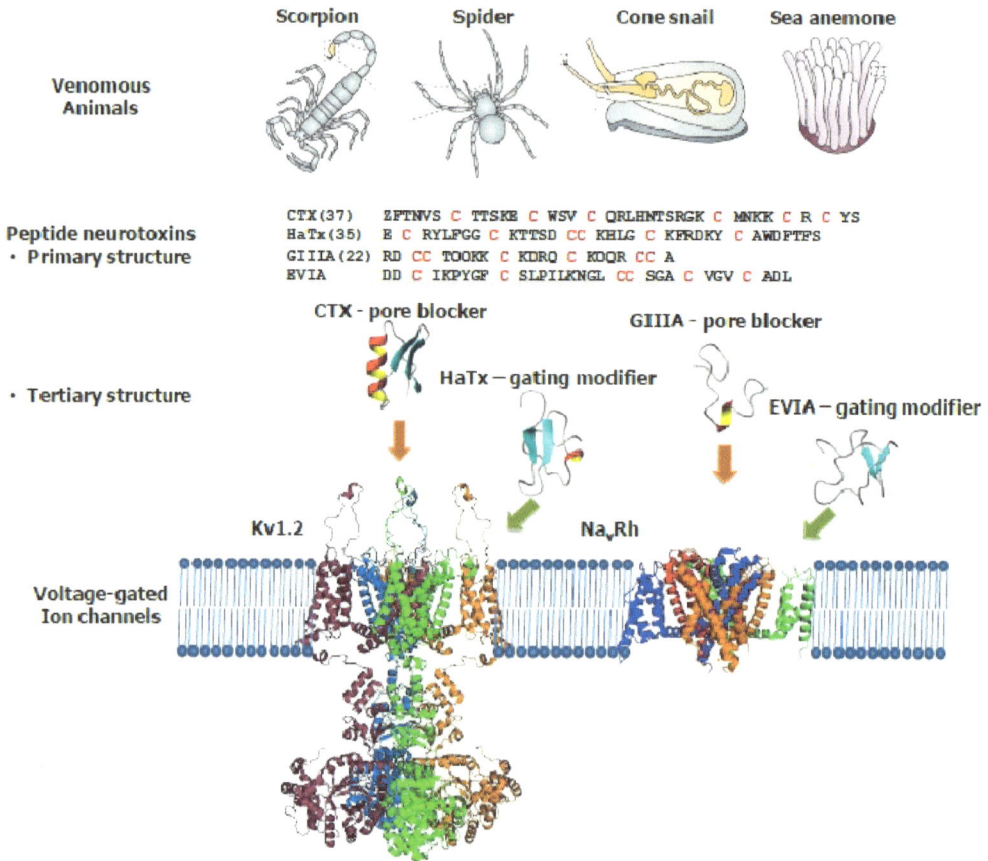

**Figure 1:** Overview of peptide neurotoxins blocking voltage gated ion channels. Peptide neurotoxins isolated from venomous animals (scorpion, spider, cone snail, and sea anemone) have three dimensional structures consisting of α-helix, β-sheet, and loops by multiple disulfide bonds (red in sequences). Peptide neurotoxin structures fit into binding sites of voltage gated ion channel subtypes. Depending on the binding sites and action mechanisms, peptide neurotoxins are classified into pore blockers and gating modifiers. The structure of Kv1.2, NavRh, charybdotoxin (CTX), hanatoxin (HaTx), GIIIA, and EVIA were obtained from PDB entries 3LUT, 4DXW, 2CRD, 1D1H, 1TCJ, and 1G1P, respectively. The animal pictures were reproduced with permission from Lewis *et al.* [11].

Other than scorpion toxins, many peptide neurotoxins from cone snails, snake, and sea anemones have been found to occlude the pore of Kv channels (Table **2**). In cone snails, κ-PVIIA and κM-RIIIK are representative pore blockers isolated from *Conus purpurascens* and *Conus radiates* [46, 47]. κ-PVIIA, a 27 amino acid toxin, is the first conotoxin to target Kv channel pores. It blocks *Shaker* channel pore but not those of splicing variants of spiny lobster (*Panulirus interruptus*) *Shaker* channel or rat brain Kv1.1 [48, 49]. Both charybdotoxin and κ-PVIIA interact with the *Shaker* pore region. However, the binding mechanisms of the two are slightly different. Phe425Gly *Shaker* mutation had increased affinity to charybdotoxin while it lost that to κ-PVIIA. Furthermore, κ-PVIIA bound to the Kv in a state dependent manner [49, 50]. κM-RIIIK, a 24 amino acid toxin, exhibits its target specificity for *Shaker* channels and human Kv1.2, with a state dependence ($IC_{50}$ = 200 nM at resting membrane potential *vs.* 400 nM at 0 mV). κM-RIIIK does not act on other Kv channels (Kv1.1, 1.3-1.6) [46, 51]. Conkunitzin-S1, a 60 amino acid toxin obtained from *Conus striatus*, also inhibits *Shaker* channel. The Lys427 mutation in the pore region increased the affinity, indicating that the toxin interacts with the pore [52]. Sea anemones also produce diverse peptide neurotoxins, which are categorized into types 1 to 4. BgK and ShK, type 1 toxins, which have been isolated from *Bunodosoma granulifera* and *Stichodactyla helianthus*, bind to the pore region of Kv1 and Kv3 [19, 53-56]. On the other hand, Type 3 toxins are classified into gating modifier toxins, while type 4 toxins have recently been isolated and their function is to be determined [20].

**Table 2:** Peptide neurotoxins for Kv channels

| Toxin name | Source | Sequence | Target Kv | Potential effects |
|---|---|---|---|---|
| Noxiustoxin | *Centruroides noxius* | TIINVK C TSPKQ C SKP C KELYGSSAGAK C MNGK C K C YNN | Kv1.2-1.3 inhibition | Multiple sclerosis or seizure (predicted) |
| Charybodotox in | *Leiurus quinquestriat us hebraeus* | ZFTNVS C TTSKE C WSV C QRLHNTSRGK C MN KK C R C YS | Kv1.3 inhibition | Multiple sclerosis, decrease in T-cell proliferation |
| Agitoxin II | *Leiurus quinquestriat us hebraeus* | GVPINVS C TGSPQ C IKP C KDAGMRFGK C MNRK C H C TPK | Kv1 inhibition | Multiple sclerosis (predicted) |
| κ-PVIIA | *Conus purpurascens* | C RIPNQK C FQHLDD CC SRK C NRFNK C V* | Shaker channel inhibition | cardiac infarction |
| κM-RIIIK | *Conus radiates* | LPS CC SLNLRL C PVPA C KRNP CC T* | Kv1.2-Kv1.7 heterotetram er inhibition | cardiac infarction |

*Table 2: contd….*

| κM-RIIIJ | *Conus radiates* | LOO CC TOOKKH C OAOA C KYKO CC KS | Kv1.2 inhibition | Seizure (predicted) |
|---|---|---|---|---|
| Conkunitzin-S1 | *Conus striatus* | KDRPSL C DLPADSGSGTKAEKRIYYNSA RKQ C LRFDYTGQGGNENNFRRTYD C QRT C LYT | Kv1.7 inhibition | Glucose-dependent insulin secretion |
| BgK | *Bunodosoma granulifera* | V C RDWFKETA C RHAKSLGN C RTSQKYRAN C AKT C EL C | Kv1.1~`.3 inhibition | Brain damage in experimental autoimmune encephalomyeli tis |
| ShK | *Stichodactyla helianthus* | RS C IDTIPKSR C TAFQ C KHSMKYRLSF C RKT C GT C | Kv1.3 inhibition | Multiple sclerosis |
| Hanatoxin | *Grammostola spatulata* | E C RYLFGG C KTTSD CC KHLG C KFRDKY C AWDFTFS | Kv2.1 inhibition | Glucose-dependent insulin secretion |
| SGTx1 | *Scodra griseipes* | T C RYLFGG C KTTAD CC KHLA C RSDGKY C AWDGTF | Kv2.1 inhibition | Glucose-dependent insulin secretion (predicted) |
| GxTX-1E | *Plesiophrictus guangxiensis* | EGE C GGFWWK C GSGKPA CC PKYV C SPKWGL C NFPMP | Kv2.1 inhibition | Glucose-dependent insulin secretion |
| JzTX-III | *Chilobrachys jingzhao* | DGE C GGFWWK C GRGKPP CC KGYA C SKTWGW C AVEAP | Kv2.1 inhibition | Glucose-dependent insulin secretion (predicted) |
| BDS-I | *Anemonia sulcata* | AAP C F C SGKPGRGDLWILRGT C PGGYGYTSN C YKWPNI CC YPH | Kv3.4 | Cell death in Alzheimer's disease models |
| BDS-II | *Anemonia sulcata* | AAP C F C PGKPDRGDLWILRGT C PGGYGYTSN C YKWPNI CC YPH | Kv3.4 | |
| APETx1 | *Anthopleura elegantissima* | GTT C Y C GKTIGIYWFGTKT C PSNRGYTGS C GYFLGI CC YPVD | Kv11.1 inhibition | Long QT syndrome |
| PIVE | *Conus purpurascens* | D CC GVKLEM C HP C L C DNS C K NYGK* | Subtype to be determined | Not known |
| PIVF | *Conus purpurascens* | D CC GVKLEM C HP C L C DNS C KKSGK* | Subtype to be determined | Not known |

*Table 2: contd....*

| Contryphan-Vn | *Conus ventricosus* | GD C PWKPW C* | Not known | Not known |
| pl14a | *Conus planorbis* | FPRPRI C NLA C RAGIGHKYPF C H C R* | Kv1.6 inhibition | Not known |
| Z, pyroglutamate; O, hydroxyproline; *, -NH2; background colors (pale blue, pink, and pale green) represent pore blockers, gating modifiers, and others, respectively. | | | | |

## Gating Modifiers

Gating modifiers comprise the second class of peptide neurotoxins for Kv channels, which bind to the voltage sensing domain and subsequently affect the gating of channels by stabilizing closed, open, or inactivated conformations (Table 1). Hanatoxin, a tarantula toxin from *Grammostola spatulata*, is the first peptide neurotoxin that has been chemically and functionally characterized as a gating modifier toxin [57-59]. It has been described that the channel bound with toxin opens with showing slow activation and fast deactivation. It has been found that one toxin peptide binds to each of four voltage sensor domains of a single tetrameric Kv channel, indicating that four toxins bind to one tetramer channel complex. The Kv channel mutagenesis studies showed that the toxins bind to S3b-S4 region which is known as the voltage sensor paddle [60]. The gating modifier toxins have been known to be partitioned into membrane and then laterally interact with sensor domains [16]. This fact is consistent with the crystallographic observation that the voltage sensor paddle is located outside and contact the membrane [28]. With producing such firm evidence from biophysics, structural chemistry and molecular biology, the class of gating modifiers for Kv channels has been being established.

Hanatoxin had appeared to be a practical tool for exploring mechanisms of gating modifier. Unfortunately, however, it is not used due to difficulties in its chemical synthesis. Instead, SGTx1, a highly related but low affinity tarantula toxin, became an alternative for mutagenesis studies to know important residues in binding [61]. A mutagenesis study and structure of SGTx1 demonstrated that important residues for Kv2.1 inhibition are clustered in one face of toxin and contain hydrophobic residues and that charged residues surround it [62]. GxTX-1E which is obtained from Chinese earth tiger tarantula *Plesiophrictus guangxiensis* is shown to be a high affinity toxin for Kv2.1 by Merck [63]. The nuclear magnetic resonance (NMR) structure of this toxin has also been determined [64]. The toxin will drive future studies on toxin-Kv channel interaction because it is relatively easy to synthesize and has a nanomolar affinity for Kv2.1, which now seems to be enough to economize the investigational approaches. For example, the double mutant cycle analysis, a method that has been useful for explaining the action mechanism of pore blockers onto Kv channels but that needs potent ligands, can be applied to gating modifier toxin mechanism with GxTX-1E [63]. Another possible contribution of GxTX-1E is Kv structuralization. In general, X-ray crystals of Kv channels are obtained when it opens because of a technical limitation which is that crystallization should be carried out at 0 mV (which looks like a neutral voltage but in fact is a depolarized one for the channel). GxTX-1E is able to stabilize a closed conformation state. Therefore, the toxin-bound form will unveil the closed channel structure, which will give a rare opportunity to understand the transition between the open and close conformations. An interesting chemical property of GxTX-1E is that the toxin is more homologous with JzTX-III [65, 66], and also with GsMTx-4 [67] than with hanatoxin or SGTx1, the Kv gating modifiers mentioned above. JzTX-III is a Nav1.5 and Kv2.1 gating modifier and GsMTx-4 blocks native mechanosenstive channels and canonical type transient receptor potential (TRPC) ion channels. On the other hand, GsMTx-4 activates ankyrin type transient receptor potential ion channel (TRPA1) [68]. Those less voltage-dependent TRP channel types are all known to play a role in sensory nervous system including pain pathways. This might be an important hint at both scientific and clinical aspects. As well as genetical similarity, functional similarity is possibly important to define voltage-dependent gating throughout the channel families. Otherwise, channels might share key structural components related to the toxin binding and those may govern the gating for a certain number of heterogeneous ion channels following a unified rule. The possible presence of common pharmacophore that determines the toxin binding likely accelerates optimization of a therapeutic ligand but at the same time, this could raise a possibility of off-target effects through the action on unwanted ion channels. Further studies are required to examine these hypotheses.

Gating modifier toxins have also been useful to obtain the structural view for the voltage sensor domain, as scorpion toxins for the pore domain [15]. The X-ray crystals of bacterial KvAP channel suggested the voltage sensor paddle (S3b-S4) model in which the sensor paddle of KvAP contacts membrane lipids and the paddle moves distantly (~15Å) toward outer membrane layer in response to membrane depolarization [27]. Toxin accessibility analysis using quenching approaches of hantatoxin's tryptophan (Trp) fluorescence revealed that S4 in resting channels only moves inwardly by ~8Å [29]. This result together with other previous consistent observations raised a possibility that the model should be corrected, highlighting these data are complementary [15].

Sea anemone Type III toxins are also known to act as a gating modifier. BDS-I and BDS-II, 43 amino acid toxins isolated from *Anemonia sulcata*, block the rapidly inactivating Kv3.4 channel but have no effect on other Kv channels and Nav channels [69]. However, recently it has been shown that BDS-I also inhibits fast inactivation of human Nav1.7 like scorpion α-toxins and δ-conotoxins (those will be discussed below) [70]. APETx1, a 42 amino acid anemone toxin, has been isolated from *Anthopleura elegantissima*. Although the toxin has a 53% homology with BDS-I, it only acts on human ether-a-go-go-related gene (hERG) $K^+$ channel, which is an essential anti-target on drug development processes since many approved drugs were withdrawn due to fatal long QT arrhythmia by hERG inhibition [71]. The toxin application shifts the channel activation curve to positive direction like hanatoxin. It does not compete with BeKm-1, a pore-blocking peptide toxin. Further, mutations in the S3b region of hERG abolished the responsiveness of APETx1, indicating that the toxin is a gating modifier [72].

### Other Toxins

Many conotoxins are waiting for their fine characterizations in terms of Kv channel inhibition. κA-conotoxins PIVE and PVIF isolated from *Conus purpurascens* promote action potential firing. Although these toxins have been suggested to act on Kv channels, their binding mechanisms and target-subtypes remain to be determined [73]. Another conotoxin contryphan-Vn is a D-Trp containing short peptide isolated from *Conus ventricosus*. It has been suggested to interact with the pore of Kv channels. However, its mechanism and target also need to be determined [74]. pl14a, a 25 amino acid toxin isolated from *Conus planorbis*, inhibits Kv1.6 as well as nicotinic acetylcholine receptor subtype α3β4 with micromolar affinities [75].

## Molecular Properties of Neurotoxin Targeting Kv Channels

All of peptide neurotoxins have six cysteines except for contryphan-Vn, pl14a, and conkunitzin-S1, which have two or four cysteines (Table **2**). Disulfide bonds by these cysteine residues are a central factor to form rigid three-dimensional structures, which enables toxins to stably fit binding pockets of voltage-gated ion channels including Kv channels. The opening of the disulfide bonds may cause loss of function. Sequence variations in inter-cysteine residues are another important factor because their intrusion into the lining of the putative pocket determines target specificity and potency. NMR has been a powerful method to define structures of such small unusual peptides, which are typically difficult to crystallize. In fact, most of structures of short peptide neurotoxins were determined by NMR. X-ray has been used for several small peptide neurotoxin such as α-conotoxins and large peptide neurotoxins like conkunitzin-S1 [76-78].

The structures of peptide neurotoxins for Kv channels indicate that folds consist of a combination of β-strands and α-helices (Fig. **2**) [14]. Mostly, the toxin structures can be categorized into two scaffolds, cystine-stabilized α/β (CSα/β) and inhibitory cystine knot (ICK) motifs [79, 80]. In CSα/β motif, an α helix is connected with a β sheet by two disulfide bonds. On the other hand, in ICK motif, the third disulfide passes through a ring formed by the first 2 bonds (1-4 and 2-5) and the intervening polypeptide backbone. In general, either of rigid scaffolds is functionally required. Interestingly, it is not true that a structural motif always correlates with a particular binding mechanism. For example, two major pore blockers, charybdotoxin and κ-PVIIA, have different scaffolds: CSα/β motif for charybdotoxin and ICK motif for κ-PVIIA, respectively [79, 81]. Also, κ-PVIIA and GxTX-1E have a same scaffold ICK motif, but they bind to different regions, pore vestibule and voltage sensing domain, respectively [64]. These facts suggest that one needs to narrow key molecular determinants down to smaller than the scaffolds for the pharmacological activity. In fact, it has been shown that pore blockers with different scaffolds have a convergent mechanism. For Kv channels, a molecular dyad is important for blocking the pore, which is composed of Lys residue that plugs the pore vestibule and a hydrophobic residues (Tyr, Phe, or Leu) separated by ~7Å [82]. For example, Lys7 and Phe9 of κ-PVIIA ICK motif have a similar spatial location with Lys27 and Tyr36 of charybdotoxin CSα/β motif [50, 79, 81]. These structure and function relationships indicate a convergent pore blocking mechanism for Kv channels. For some other toxins such as κ-PVIIA, BgK and dendrotoxin-I, two surrounding hydrophobic residues are required along with an essential Lys, which indicates, in this case, a triad [83]. Sea anemone toxins ShK and BgK are other good examples of convergent mechanisms [82]. ShK adopts a helical-cross like motif in which two α helices are located perpendicular to each other, while BgK adopts a helical capping motif in which one α helix caps the other two. Despite the different scaffolds, functional dyads, Lys22 and Tyr23 of ShK and Lys25 and Tyr26 of BgK, are conserved [84].

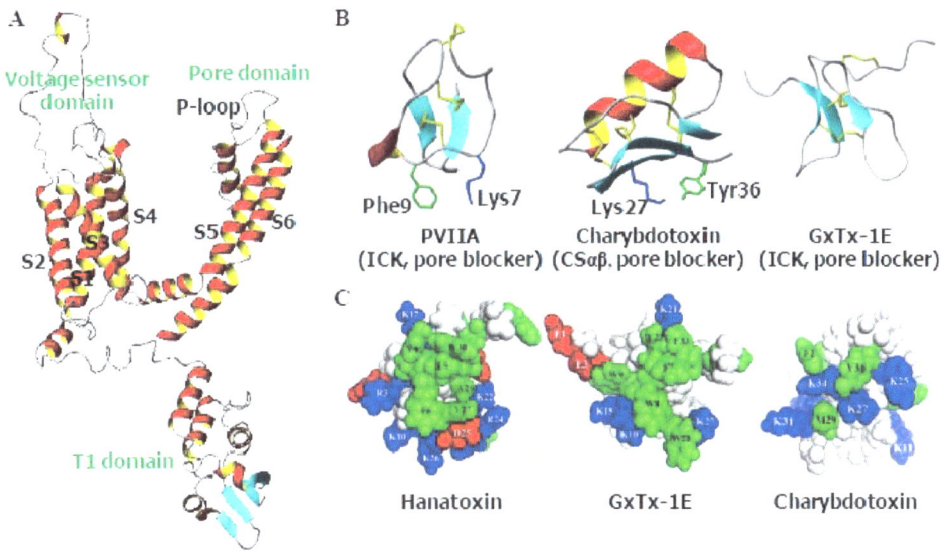

**Figure 2:** Three dimensional structures of the monomer of Kv1.2 and peptide neurotoxins for Kv channels. (A) The ribbon structure of Kv1.2 monomer shows its voltage sensor domain (consisting of S1 to S4), pore domain (consisting of S5, P-loop, and S6), and intracellular T1 domain. Peptide neurotoxins bind to extracellular parts of voltage sensor or pore domains. (B) Ribbon structures of PVIIA, charybdotoxin, and GxTx-1E. The structures of PVIIA and GxTx-1E adopt the inhibitory cystine knot motif, while that of charybdotoxin adopts the cysteine-stabilized α-helix and β-sheet motif. The hydrophobic and positively charged residues of functional dyads of PVIIA (Lys7 and Phe9) and charybdotoxin (Lys27 and Tyr36) are colored by green and blue, respectively. The disulfide bonds are colored by yellow. (C) Space filling models of hanatoxin, GxTx-1E, and charybdotoxin. Hydrophobic, basic, and acidic residues are colored by green, blue, and red, respectively. Structures in (C) were modified forms of those in Lee *et al.* with permission from [64]. The structures of Kv1.2, PVIIA, charybdotoxin, and GxTx-1E were obtained from PDB entries 3LUT, 1AV3, 2CRD, and 2WH9, respectively.

Most of gating modifier structures adopt an ICK motif and consist of a triple-stranded anti-parallel β sheet connected by four reversals and disordered N- and C-terminal segments. In some toxins, β strand I is substituted with $3_{10}$ helix (Fig. **2**). The three dimensional structures of hanatoxin and GxTX-1E, the gating modifiers for Kv2.1, implicate a hydrophobic patch surrounded with charged residues, showing amphipathic characteristics. Alanine scanning mutagenesis of SGTx1 has shown that Leu5, Phe6, and Trp30 are critical for pharmacological function and are located in the hydrophobic patch [85]. It is likely that the hydrophobic patch interacts with hydrophobic residues of channels or the membrane bilayer. The hydrophobic patch of gating modifier may also play a role in toxin's partitioning into the membrane to access the membrane spanning voltage sensor [16]. The folding and amphipathic features are common for structures of gating modifier neurotoxins, but the detailed differences in hydrophobic patches seem to be responsible for the individual potency and selectivity.

**Therapeutic Implications**

As for therapeutics, a pore blocker toxin κ-PVIIA (CGX-1051 from Cognetix Inc.) from *Conus purpurascens* has been examined in preclinical study for myocardial infarction. Intravenous injection of κ-PVIIA attenuated cardiac infarction in animals [47, 86, 87]. Bolus injections of 100 and 300 µg/kg κ-PVIIA into dogs and rats after the onset of ischemia significantly reduced infarct size. Bolus injection of 10 and 100 µg/kg κ-PVIIA showed same effect in the rabbit. Interestingly, κ-PVIIA did inhibit *Shaker* channels but not rat Kv1.1 as mentioned above. Thus, the beneficial effect is promising but the molecular mechanism remains to be explored. It would be interesting to determine whether such a strategy using Kv pore blockers or other administration routes that skip the blood-brain barrier can improve cerebral ischemic injuries.

Another pore blocker for Kv1.2, κM-RIIIK, also has cardioprotective effects at 100 µg/kg intravenous bolus injection in rats [88]. Interestingly, a more potent blocker for Kv 1.2, κM-RIIIJ, did not exhibit cardioprotective effects in rats, although the toxin possesses 10-fold high affinity compared to κM-RIIIK. Rather, κM-RIIIK has shown high potency in blocking Kv1.2-Kv1.7 heterotetramer compared to κM-RIIIJ. The results indicate that κM-RIIIK may give cardioprotection *via* blocking heterotetramer of Kv channels. At a mechanistic angle, however, these data should also be carefully translated. Since $K^+$ channel blockade will result in membrane depolarization of excitable cells and in turn increases in excitotoxicity by elevating intracellular $Ca^{2+}$ level, $K^+$ channel blockers are not typically believed to be cytoprotective. Although the neurotoxins blunt the peak current through Kv channels, it would be state-dependent or affect state-transitions. For instance, the toxin block might lead to an increase in the $K^+$ current at the bottom line by preventing Kv channels from entering a certain less active state. These $K^+$ currents may render excitable cells repolarized or hyperpolarized, less excitable, relaxed, and finally more surviving. This hypothesis is waiting to be explored [88].

ShK-192 is a derivative of an anemone toxin ShK with an improved chemical stability against changes in pH. Despite its cellular target not being a neuronal type, ShK-192 is aiming at reversing a disease state of the neural system, multiple sclerosis. ShK-192 is under clinical trials by Airmid Inc. and Kineta, Inc. [89, 90]. ShK-192 contains a nonhydrolyzable phosphotyrosine surrogate, a methionine isostere, and a C-terminal amide. Nevertheless, its overall scaffold is similar to that of ShK. Also, ShK-192 retains its specificity and potency for Kv1.3. Even though the current developments of peptide neurotoxins κ-PVIIA and ShK-192 are not aiming at CNS diseases, these pharmacological results and developmental status in other clinical fields shed lights on the potential therapeutic usefulness of peptide toxins for targeting CNS diseases.

Many CNS diseases can be improved by modulations of Kv channel subtypes [25]. The activation or disinactivation of Kv1.1 reduces seizure [91]. Kv1.1-ablated mice and Kv1.2-ablated mice both have shown seizures and structural changes in CNS [92, 93]. In fact, human episodic ataxia and myokymia were reported to associate with Kv1.1 mutations [94]. Even with normal functions, mild redistributions of these channels owing to neuronal injuries might also be involved in changes in electrical conduction [95, 96].Thus, strategies using activators or disinactivators for Kv1.1 and Kv1.2 channels might be beneficial in this regard.

Amyloid peptide elevated in Alzheimer's disease increases Kv3.4's expression and activity, which is potentially involved in disease exacerbation [97]. Thus Kv3.4 blockers have been suggested to blunt the progress of Alzheimer's disease.

M-current that is named because the muscarinic acetylcholine receptor activation typically couples with is known to be, at least partly, mediated *via* activation of Kv7 channels. These Kvs also act as effectors for other G-protein coupled receptor-mediated inhibitory functions in nervous system. Therefore, defects in the channel function often lead to CNS diseases such as epilepsy and cognitive disorders. Indeed, the activation of Kv7.2/7.3 has beneficial effects [98]. Retigabine has been a leading small molecule, and finally is approved by US FDA and European Medicines Agency in 2011. The drug displayed anticonvulsant activity in an extended forms of seizures during preclinical and clinical studies and is currently prescribed for partial onset seizure. Interestingly, post-herpetic neuralgia was another potential indication in its phase II clinical study. Although it failed to show significant improvement of the disease state, it suggests that pharmacology reversely stresses a pathophysiological role of the Kv [99-101]. Other possible involvement in psychiatric disorders has been suggested including anxiety, bipolar disease, schizophrenia, attention deficit–hyperactivity disorder, *etc.* [102, 103]. It is interesting whether peptide neurotoxins give an impression in reversing those pathologic CNS states.

## VOLTAGE-GATED SODIUM CHANNELS (NAV CHANNELS)

### Molecular Properties of Nav Channels

Nav channels are responsible for action potential generation and propagation in neurons and muscle cells. Nav channels open in response to depolarized membrane potentials, selectively permeating sodium ion ($Na^+$) into cells and then are quickly inactivated. As a result, action potential is generated and the channels specifically contribute to forming the rising phase of action potentials. Consecutive activations of adjacent Nav channels, affected by the initial depolarization enable spatial movement of action potentials. Pathologic functions of Nav channels likely cause abnormal absence of the action potential or excessive depolarization. Their functional significance in nervous system has been verified by many genetic diseases such as periodic paralysis, epilepsy, migraine, and chronic pain [104, 105]. Individual Nav channels consist of an α subunit that conducts sodium ion, and one or two auxiliary β subunits that modulate the gating kinetics and voltage dependence of channel activation and inactivation. Ten α subunit subtypes (Nav1.1-1.9 and NaX) and four β subunits have been identified [106, 107]. Each subtype has different tissue expression profiles. For example, Nav1.1, 1.2, 1.3, and 1.6 are expressed in CNS, while Nav1.7, 1.8, and 1.9 are expressed in PNS. Nav1.4 and Nav1.5 are expressed in skeletal muscle and cardiac muscle, respectively. NaX is expressed in cardiac and skeletal muscles and sensory neurons but its functional characteristics are poorly understood [108]. Often, Nav channels are categorized according to the extent how much a famous (or notorious) fugu toxin, tetrodotoxin (TTX) is required for blocking the channels, so called, TTX-sensitive and TTX-resistant Navs. Roughly, less sensitive channels have been known to be responsible for conductions of a specific subset of peripheral nerve fibers relaying painful signals [109]. Thus, this toxin is pharmacologically useful for discerning the related sensory neurons in pain studies. For example, very recently, identification of nociceptors and Nav subtypes were performed with TTX in characterizing a novel pain-producing endogenous mediator in diabetic neuropathy [110]. Nav 1.5, Nav1.8, and Nav1.9 are TTX-resistant channels, as they respond to higher than $\sim\mu M$ TTX. The other Navs are sensitive to nanomolar TTX [106].

The α1 subunits of Nav channels have four homologous domains with six transmembrane segments in each, which is equivalent to a subunit of Kv channels. Like Kv channels, S1-S4 segments in each domain serve as voltage sensor, and S5, P-loop, and S6 form the pore that is in charge of ion conductance and selectivity. Very recently, the crystals of bacterial Nav channels, NavAb and NavRh, have been obtained so that it is now possible to structurally delineate details (Fig. **4**) [111, 112]. Overall structures of NavAb and NavRh are very similar with those of Kv channels, but the detailed comparisons showed several differences. Pore-loops consist of two pore helices P1 and P2 differently from the pore structure of Kv channels. Also, the structure has lateral openings leading from membrane to the lumen of the pore, indicating that lipid and hydrophobic drug can bind to the pore through the pathway. In voltage sensing domain, S4 segments of each subunit (for human Navs, of each domain) have four to eight positively charged residues flanked by two hydrophobic residues. The S4 segments are located most outside from the pore region, implicating that the structure is open channel conformation like those of Kv channels. However, S3 segment of the Nav is a single straight α-helix with loose S3-S4 loops, while Kv channels have two helices S3a and S3b. Future studies on structures will keep forming better homology models that push the limit of virtual predictions of pharmacophore and structure-based drug designs.

A number of synthetic blockers for Nav channels are clinically popular. Therapeutic purposes of most of those Nav channel drugs are local anesthesia and antiarrythmia and some of blockers are being used for the both purposes with different administration forms and regimens. Examples are lidocaine, bupivacaine, mepivacaine, prilocaine, benzocaine, novocaine, *etc.* Unfortunately, their utility is limited due to side effects that originate from their less Nav specificity. For instance, local anesthetics like procaine might cause arrhythmia when high doses are accidentally administered or when not properly localized [113]. This is because Nav channels of other subtypes or of other neighboring tissues are blocked. In fact, the limitation for their use in which systemic administration is only carried out for particular indications, resulted from this risk.

A unique approach to overcome such less specificity of Nav blockers was reported by Clifford Woolf and his colleagues in 2007 [114]. QX-314 is a quaternary derivative of the local anesthetic lidocaine. It is first synthesized by Astra Pharmaceutical Company in early 1970s [115]. Its less membrane permeability caused by positively-charged ammonium moiety, prevents from accessing common cytosolic Nav binding sites for lidocaine species at a critical amount to block the channel. This molecular property has limited its use to an experimental one for basic Nav channel studies [116]. The Woolf group found that QX-314 can be transported through transient receptor potential vanilloid subtype 1 (TRPV1) which is a nociceptive sensory neuron-specific ion channel. When TRPV1 is open by its specific agonist, QX-314 which stays outside tissues, eventually goes into the cytosolic area of the pain-specific neurons through the TRPV1 pore, which leads to Nav channel blockade by it only in the nociceptive neurons (Fig. **3A**). Therefore, surprisingly, specifically of the nociceptors, excitation and conduction *via* Nav channels can be silenced. This result suggested that QX-314, a lidocaine derivative that was therapeutically useless due to its poor target-accessibility, may resurrect as a local analgesic when combined with a specific TRPV1 agonist. This is extremely unique because side effects such as sensory loss of other innocuous feelings, motor defects, or arrhythmia by blocking Nav channels in non-nociceptive sensory neurons, motor neurons, or cardiac tissues could be less predictable than when current lidocaine-like anesthetics are administrated. The same group is trying to optimize the combination-ratio of the drugs and examine which pain types are effectively controllable by this strategy. To synthesize new analogs which are more permeable to TRPV1 but less membrane permeable might also be a better future option. This strategy might extend their indications beyond pain control. TRPV1-expressing nerve fibers not only take part in pain mediation but also in diseases caused by neurogenic inflammation including asthma, pancreatitis, irritable bowel syndrome, psoriasis *etc.* [117]. Thus, it would be interesting to observe whether the hypothesis of this pharmacologically-accomplished Nav specificity works for such diverse diseases in the near future.

Recently, subtype-specific Nav channel blockers have been developed for same purpose [118]. Different from cases like above pharmacological trimming of old drugs, straightforward approaches to achieve Nav specificities might generate information throughout Nav subtypes, which does not depend on other molecular partners and readily obtains CNS therapeutic candidates [119]. Toxin interactions and related structural information discovered so far should be helpful on this avenue.

**Peptide Neurotoxins for Nav Channels**

The interactions between Nav channels and the peptide neurotoxins provide valuable information for Nav structure as well as opportunities for therapeutic development. Seven toxin-binding sites (site1-7) are established in Nav channels. Through binding to the sites, toxins affect ion permeation and gating properties (Fig. **3B**) [120]. Peptide neurotoxins, mainly isolated from sea anemones, scorpions, spiders, and cone snails, act at sites1, site3, site4, and site6 [120-122]. Site2, site5, and site7 are for alkaloid toxins, polyether toxins, and pyrethroids, respectively, all of which are non-peptidal neurotoxins. Tetrodotoxin (TTX), a guanidium alkaloid from fugu bacteria, is a representative non-peptidal toxin that binds to pore regions (site1) with a high affinity (~nM) to TTX-sensitive Nav channels.

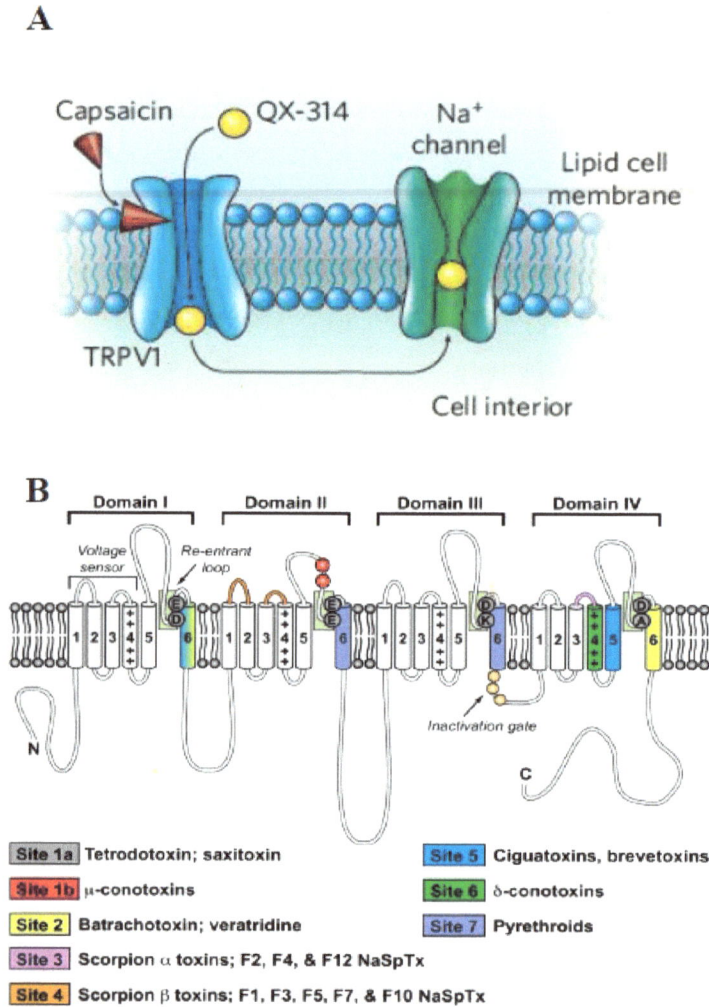

**Figure 3:** the principle of the QX-314 action mechanism and toxin binding sites of Nav channels. (A) QX-314 can penetrate the cellular membrane of the nociceptors through the pore of transient receptor potential vanilloid subtype 1 (TRPV1), which is a nociceptor-specific ion channel. Upon TRPV1 pore opening by its activators (*e.g.,* capsaicin in this figure), QX-314 which stays outside tissues, eventually goes into the cytosolic area of the pain-specific neurons through the TRPV1 pore. Intracellular QX-314 binds to its Nav binding site, which leads to Nav channel blockade. Since it only occurs in the nociceptive neurons, Adverse effects *via* non-nociceptive neurons or motoneurons are less possible than other hydrophobic local anesthetics. The figure is reproduced with permission from [303]. (B) The seven toxin binding sites on Nav channels. The figure is reproduced with permission from Klint *et al.* [120].

## Pore Blockers

Although Nav pore modulation is less common in cases of peptide neurotoxins, conotoxins were found to act on the pore. μ-conotoxins isolated from the venom of cone snails consist of 16-26 amino acid residues and have six cysteines (-CC-C-C-CC-) forming three disulfide bonds. In 1985, GIIIA, GIIIB, and GIIIC from *Conus geographus* were first reported among μ-conotoxins and those block muscle Nav channels [123]. Since then many μ-conotoxins have been isolated to have 18 μ-conotoxins. A binding competition study has shown that these toxins inhibit sodium currents by binding to site1 like TTX [124]. However, their binding sites seem to be only partially overlapping according to the observation that some mutations affecting TTX binding did not change μ-conotoxin binding [125]. Based on molecular size, μ-conotoxins may bind to a wider region than TTX does. In addition, by using μ-conotoxin pharmacology, Nav subtypes can be discriminated, such as muscle type Nav1.4 and neuronal type Nav1.2 because μ-conotoxins also bind to TTX-resistant Navs and different μ-conotoxins cover different suptypes [126]. Mutagenesis of Nav channels revealed that μ-conotoxin GIIIA binds to S5-S6 of domain II of the muscle type Nav1.4 [127, 128]. KIIIA, another 16 amino acid μ-conotoxin, is isolated from *Conus kinoshitai*. KIIIA blocks a neuronal subtype Nav1.2 [129]. It has been shown that KIIIA and TTX bind to the pore together at a time [130, 131]. Interestingly, conotoxin GS, a 34 amino acid toxin, compete with TTX and GIIIA, indicating that the toxin bind to the pore region, although cysteine pattern of conotoxin GS (-C-C-CC-C-C-) is different from that of GIIIA [132]. Recently, Tx1, a large spider toxin from *Phoneutria nigriventer* was characterized to compete with μ-conotoxins and showed state-dependent binding [133].

## Gating Modifiers

Typically peptide neurotoxin binding sites (site3, site4, and site6) are scattered on voltage sensing domain. The gating modifiers that bind to these regions can differentially modulate the activation and inactivation of Nav channels. Scorpion α-toxins, consisting of 60-80 amino acids and four disulfide bonds, inhibit fast inactivation by binding to site3 located in loop between S3 and S4 in domain IV [134]. Aah II, a representative scorpion α-toxin isolated from *Androctonus australis* Hector, removes fast inactivation of mammalian Nav channels, resulting in prolonged action potentials [135]. Interestingly, it has been shown that another α-toxin Lqh III from *Leiurus quinquestriatus* Hebraeus enhanced slow inactivation that is related to a conformational change different from that of fast inactivation [136]. δ-conotoxins isolated from the venom of cone snails have similar functions with that of the scorpion α-toxins which usually inhibit Nav channel inactivation. The binding site of δ-conotoxins has been defined as site6 located in S4 in domain IV, which is different from but close to that of scorpion α-toxins [137]. Even some scorpion α-toxins and δ-conotoxins are reported to share binding sites [138]. δ-EVIA isolated from *Conus ermineus* is an interesting toxin in that it specifically acts on neuronal Nav channels over muscle and cardiac Nav channels [139]. It inhibits inactivation.

Conventionally, it has been known that scorpion β-toxins, consisting of 60-80 amino acids and knotted by four disulfide bonds, shift the voltage dependence of activation to less depolarized membrane potentials by trapping voltage sensor [140]. The mutagenesis and double-mutant cycle analysis revealed that the toxins bind to site4 located in S3-S4 in domain II [141]. However, the effect is promiscuous throughout Navs and should be more explored. Tz1 isolated from *Tityus zulianus* was shown to induce enhanced channel activation for Nav1.4 and channel inhibition for Nav1.5 by interaction with voltage sensor of domain II [142]. The same group showed that the subtype specificity of Tz1 has been also determined by the pore loop of domain III [143]. This might result from multiple interactions due to a large size of scorpion β-toxins or an allosteric mechanism connecting the voltage sensor of domain II and the pore of domain III. Further studies will confirm this prediction.

MrVIA and MrVIB classified as μO-conotoxins have been isolated from *Conus marmoreus* and have shown to block sodium currents in *Aplysia* neurons and $Na^+$ and $Ca^{2+}$ currents in *Lymnaea* neurons [144, 145]. The same toxins block both TTX-resistant and TTX-sensitive currents of rat dorsal root ganglion (DRG) neurons at different concentrations [146]. The toxins were without effect on rat $Ca^{2+}$ currents. These toxins do not compete with TTX, saxitoxin (STX), or μ-conotoxin but do with the scorpion β-toxin Ts1, indicating that the toxins likely bind to S4 site in voltage sensing domain [147, 148]. Like scorpion β-toxin, selectivity of MrVIA for Nav1.4 over Nav1.2 is related to P-loop in domain III according to electrophysiology using domain swapping channel constructs [149]. LtVIIA and LtVIC have been characterized from *Conus litteratus* and their recombinant forms inhibit sodium currents of rat dorsal root ganglion neurons similar to MrVIA and MrVIB [150, 151]. ProTx-I and ProTx-II have been isolated from *Thrixopelma pruriens* was reported to block Nav channels [152, 153]. These toxins, like hanatoxin, a gating modifier for Kv channels, shift voltage dependence of activation to more depolarized potential.

## *Other Toxins*

μT-conotoxin LtVD, a 12 amino acid toxin, has been isolated from *Conus litteratus* and shown to inhibit TTX-sensitive sodium currents of dorsal root ganglion neurons with $IC_{50}$ of 156 nM. This is a first example of T-superfamily conotoxins (--CC--CC-; cf. conotoxins have four different superfamilies (O, A, M and T) according to disulfide patterns and prepopetide precursor sequences) that act on Nav channels [154]. Their binding site and mechanism are to be determined. Conversely, ι-conotoxins activate Nav channels *via* a different mechanism from that of δ-conotoxins. RXIA, a 46 amino acid toxin, was characterized from *Conus radiates* and shown to elicit iterating action potentials on amphibian peripheral nerves [155]. Later, the specificity of the toxin has been determined to shift the voltage-dependence of activation of Nav1.2, Nav1.6, and Nav1.7 to more hyperpolarized level [156]. LtIIIA, a 17 amino acid toxin, has been characterized from *Conus litteratus* and it belongs to M-superfamily (-CC-C-C-CC-). Unlike conventional M-superfamily conotoxins that occlude the pores of Nav and Kv channels like GIIIA and RIIIK (see above), LtIIIA has a short third inter-cysteine loop and increases TTX-sensitive currents without inhibiting inactivation [157]. The binding site and mechanisms of ι-conotoxins remains to be determined.

## Molecular Properties of Peptide Neurotoxins Targeting Nav Channels

Like peptide neurotoxins for Kv channels, the primary sequences of Nav-targeting toxins have 6 or 8 cysteines (Table **3**). μ-conotoxins, representative pore blockers for Nav channels, have three loop structure with $3_{10}$ helix in some μ-conotoxins depending on the length of inter-cysteine loop. The net charges of μ-conotoxins are positive like other pore blockers for Kv and Cav channels, suggesting the potential importance of electrostatic interactions. The positively charged residues (Arg or Lys) in inter-cysteine loop 2 of μ-conotoxins seem to be essential for their binding to Nav channels. Those residues have been suggested to plug into the vestibule of pore region in Nav channels. For example, Arg13 of GIIIA is an important residue for the action on Nav1.4 (Fig. **4**). The guanidium group of Arg is the reminiscent of that of TTX or STX. Interestingly, another pore blocker conotoxin GS adopts ICK motif and consists of anti-parallel β-sheet like other Kv channel blockers [158]. Although the scaffolds of μ-conotoxins and conotoxin GS are different, it has been suggested that positively charged residues of the two toxins have similar spatial distribution. However, the key residue for the activity has not been confirmed yet.

**Table 3:** Peptide neurotoxins for Nav channels

| Toxin name | Source | Sequence | Target Nav | Potential effect |
|---|---|---|---|---|
| GIIIA | *Conus geographus* | RD CC TOOKK C KDRQ C KOQR CC A* | Nav1.4 inhibition | Hyperkalemic periodic paralysis (predicted) |
| GIIIB | *Conus geographus* | RD CC TOORK C KDRR C KOMK CC A* | Nav1.4 inhibition | |
| GIIIC | *Conus geographus* | RD CC TOOKK C KDRR C KOLK CC A* | Nav1.4 inhibition | |
| KIIIA | *Conus kinoshitai* | CC N C SSKW C RDHSR CC* | Nav1.2 inhibition | Seizures and chronic pain (predicted) |
| Conotoxin GS | *Conus geographus* | A C SGRGSR C PPQ CC MGLR C GRGNPQK C IGAHEDV | Nav1.4 inhibition | Hyperkalemic periodic paralysis (predicted) |
| Tx1 | *Phoneutria nigriventer* | AELTS C FPVGHE C DGDASN C N CC GDDVY C G C GWGRWN C K C KVADQSYAYGI C KDVN C PNRHLWPAKV C KKP C RRN C GG | Nav1.2 inhibition | Seizures and chronic pain (predicted) |
| Aah II | *Androctonus australis Hector* | VKDGYIVDDVN C TYF C GRNAY C NEE C TKLKGESGY C QWASPYGNA C Y C YKLPDHVRTKGPGR C H | Nav potentiation | Not known |
| Lqh III | *Leiurus quinquestriatus Hebraeus* | VRDGYIAQPEN C VYH C FPGSSG C DTL C KEKGGTSGH C GFKVGHGLA C W C NALPDNVGIIVEGEK C HS* | Nav1.4 and Nav1.7 potentiation | Not known |

*Table 3: contd...*

| δ-EVIA | *Conus ermineus* | DD C IKPYGF C SLPILKNGL CC SGA C VGV C ADL* | Nav1.2 potentiation | Not known |
|---|---|---|---|---|
| Lqq IT2 | *Leiurus quinquestriatus Quinquestriatus* | DGYIRKRDG C KLS C LFGNEG C NKE C KSYGGSYGY C WTWGLA C W C EGLPDDKTWKSETNT C G | Insect Nav channel inhibition | Pain and convulsion (predicted) |
| Lqh IT2 | *Leiurus quinquestriatus Hebraeus* | DGYIKRRDG C KVA C LIGNEG C DKE C KAYGGSYGY C WTWGLA C W C EGLPDDKTWKSETNT C G | Insect Nav channel inhibition | Pain and convulsion (predicted) |
| BmK AS | *Buthus martensii Karsch* | DNGYLLDKYTG C KVW C VINNES C NSE C KIRGGYYGY C YFWKLA C F C QGARKSELWNYNTNK C NGKL | Nav1.3 inhibition | Pain and convulsion (predicted) |
| Tz1 | *Tityus zulianus* | KDGYLVGNDG C KYS C FTRPGTY C ANE C SRVKGKDGY C YAWMA C Y C YSMPNWVKTWDRATNR C GR | Nav1.4 activation Nav1.5 inhibition | Not known |
| Cn2 | *Centruroides noxius Hoffmann* | KEGYLVDKNTG C KYE C LKLGDNDY C LRE C KQQYGKGAGGY C YAFA C W C THLYEQAIVWPLPNKR C S* | Nav1.6 activation | Not known |
| MrVIA | *Conus marmoreus* | A C RKKWEY C IVPIIGFIY CC PGLI C GPFV C V | Nav1.8 inhibition | Chronic pain |
| MrVIB | *Conus marmoreus* | A C SKKWEY C IVPILGFVY CC PGLI C GPFV C V | Nav1.8 inhibition | |
| LtVIIA | *Conus litteratus* | C LGWSNY C TSHSI CC SGE C ILSY C DIW | Subtype to be determined | Not known |
| LtVIC | *Conus litteratus* | WP C KVAGSP C GLVSE CC GT C NVLRNR C V | Subtype to be determined | Not known |
| ProTx-I | *Thrixopelma pruriens* | E C RYWLGG C SAGQT CC KHLV C SRRHGW C VWDGTFS | Nav1.8 inhibition | Chronic pain (predicted) |
| ProTx-II | *Thrixopelma pruriens* | Y C QKWMWT C DSERK CC EGMV C RLWCKKKLW | Nav1.7 inhibition | Chronic pain (predicted) |
| LtVD | *Conus litteratus* | D CC PAKLL CC NP | TTX-sensitive current inhibition | Chronic pain (predicted) |

*Table 3: contd...*

| RXIA | *Conus radiates* | GOSFCKADEKO C EYHAD CC N CC LSGI C AOSTNWILPG C STSSFFKI | Nav1.6 activation | Not known |
|------|------------------|--------------------------------------------------------|-------------------|-----------|
| LtIIIA | *Conus litteratus* | Dγ CC γ OQW C DGA C D CC S | TTX-sensitive current enhancement | Not known |

Z, pyroglutamate; O, hydroxyproline; *, -NH2; γ, gammacarboxyglutamate; background colors (pale blue, pink, and pale green) represent pore blockers, gating modifiers, and others, respectively.

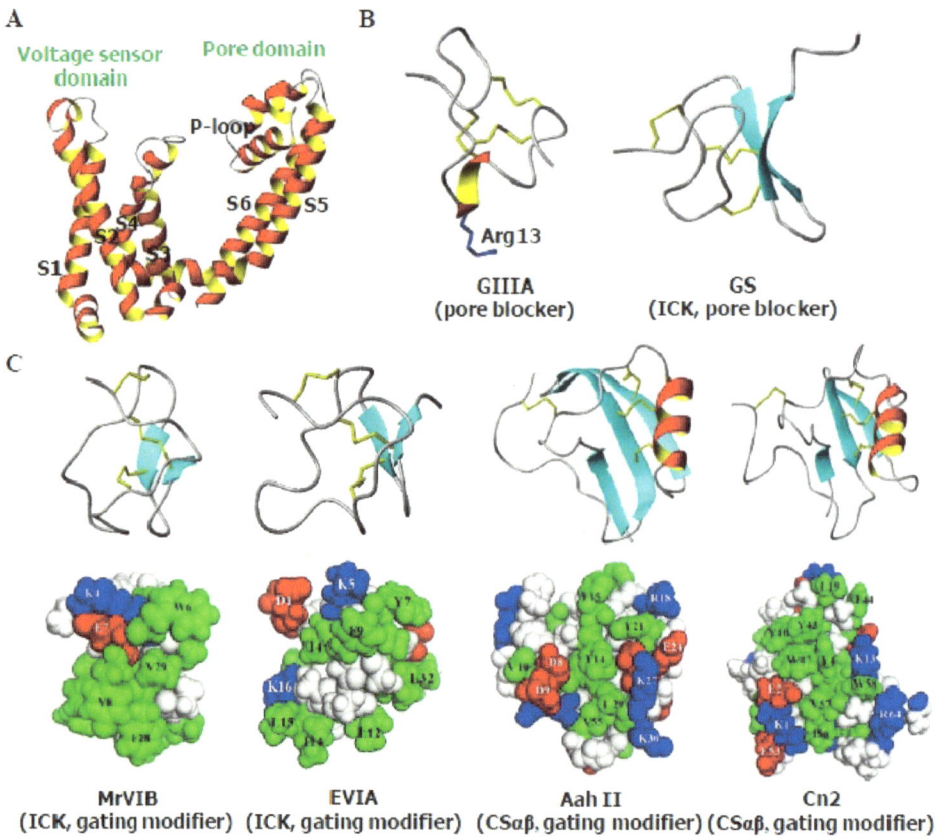

**Figure 4:** Three dimensional structures of the monomer of NavAb and peptide neurotoxins for Nav channels. (A) The ribbon structure of NavRh monomer shows its voltage sensor domain consisting of S1 to S4 and pore domain consisting of S5, P-loop, and S6. Peptide neurotoxins bind to extracellular parts of voltage sensor and pore domains. (B) Ribbon structures of two pore blockers GIIIA and GS having three loop structures and inhibitory cystine knot motifs. The important residue Arg13 is colored by blue. (C) Ribbon structures and space filling models of gating modifiers MrVIB, EVIA, Aah II, and Cn2. In space filling models, hydrophobic, basic, and acidic residues are colored by green, blue, and red, respectively. In the ribbon structure, disulfide

bonds are colored by yellow. The structures of NavAb, GIIIA, GS, MrVIB, EVIA, Aah II, and Cn2 were obtained from PDB entries 3RVY, 1TCJ, 1AG7, 1RMK, 1G1P, 1PTX, and 1CN2, respectively.

Structures of gating modifiers for Nav channels are grouped into two classes of backbone scaffolds. Scorpion toxins adopt CSαβ motif, while μO conotoxins and δ-conotoxins adopt ICK motif. For example, both structures of scorpion α-toxin Aah II and β-toxin Cn2 have βαββ-type CSαβ scaffolds (Fig. **4**). In detail, there are some differences in loop region between second and third β sheets and C-terminals [159, 160]. These differences could probably be a major molecular determinant of binding sites and even actions of α-toxin Aah II and β-toxin Cn2. As for the two gating modifiers, μO conotoxin MrVIB and δ-conotoxin EVIA, the toxins share the ICK backbone scaffold [146, 161]. Interestingly, many pore blockers also employ ICK motif, suggesting that inter-cysteine residues might be more responsible for the action mechanisms than disulfide scaffolds.

The structures of Nav gating modifiers have hydrophobic cluster on one face, which might be important for membrane accessibility. This feature would also be required for interaction with voltage sensor units since the sensor is buried in the membrane. A structural analysis of scorpion α-toxin Aah II and other α-toxins has demonstrated that those toxins conserve a hydrophobic surface and that it is important for toxin action as well as loop regions between second and third β sheets and C-termini [162]. Subtle differences in hydrophobic patch residues may finely tune the Nav specificity of each toxin. The extraction of the pharmacophore of the gating modifiers within the hydrophobic face must be useful to design toxin-mimicking specific Nav-targeting therapeutics in the near future.

**Therapeutic Implications**

Nav1.3, Nav1.7, Nav 1.8, and Nav1.9 are expressed in the sensory afferent (DRG, trigeminal, and vagal) neurons that are highly involved in pain signaling [163-165]. For instance, gain and loss of function of Nav1.7 are related to several genetic diseases such as paroxysmal extreme pain disorder, inherited erythromelalgia, and congenital indifference to experience pain [166-169]. ProTx-I and -II from a tarantula spider *Thrixopelma pruriens* has been reported to block multiple Nav channels including such pain-mediating ones [153]. Later, it has been shown that proTx-II blocked Nav 1.7 with $IC_{50}$ of 0.3 nM, while the toxin blocked other Nav channels with $IC_{50}$ of 30 to 150 nM [170]. This 100-fold selectivity for Nav1.7 may open a new possibility for specific control of pain with limited side effects [171]. The proTx-II exhibited lethality with 1 mg/kg intravenous injections and 0.1 mg/kg intrathecal injections into rats. This might be a major caveat to overcome, probably resulting from a nonselective action at higher doses. For the toxin itself, it would be conceivable whether botox-like approaches including optimizing low doses and localized routes to find a safety window can be achieved when desired neural effects maintained.

µO-conotoxins MrVIA and MrVIB from *Conus marmoreus* have had an particular interest because it is the first peptide neurotoxin blocking TTX-resistant Nav channels in DRG neurons [146]. MrVIB exhibited blocking effects with ~100 nM $IC_{50}$ both for native TTX-resistant currents of rat DRG neurons and for currents *via* heterologously expressed Nav1.8 [172]. The intrathecal MrVIB injection (0.03-3 nmol) attenuated mechanical and thermal pain in a neuropathic pain model using peripheral nerve ligation and in CFA-derived inflammatory pain model without motor deficit. Motor side effects were observed only at 30-fold higher doses. In formalin test, intrathecal MrVIB reduced secondary pain while showed no significant effect on first pain, suggesting that this toxin reduced pathological pain rather than normal sensation. The toxin expressed as a tethered form in a nociceptor specific manner showed analgesic effects [173], supporting the high possibility of chronic pain therapeutic development using the toxin. The toxin (CGX-1002 from Cognetix Inc.) is under development as pain therapeutics.

For a pore blocker conotoxin KIIIA, therapeutic possibility was also raised first in pain field. KIIIA was shown to block greater than 80% of the TTX-sensitive Nav current and about 20% of the TTX-resistant current in mouse DRG neurons [174]. Subtype overexpressing oocyte electrophysiology revealed that Nav1.2 and Nav1.6 are main targets but that Nav1.3, Nav1.4, and Nav1.5 are also blocked at nanomolar through micromolar ranges. Blockade of Nav1.2 is highly irreversible. Interestingly, KIIIA displayed *in vivo* strong anti-nociceptive activities in formalin pain test (the $ED_{50}$ was 0.1mg/kg). The second phase (where CNS synaptic processing is involved) appeared to be more. In fact, other CNS defects would be a target considering the specificity of KIIIA for Nav1.2. Since the KIIIA is among toxins that contain relatively small numbers of amino acids, further engineering to maximize its neural effects or extend therapeutic window would be easier.

## VOLTAGE-GATED CALCIUM CHANNELS (CAV CHANNELS)

### Molecular Properties of Cav Channels

Cav channels play important roles in neurotransmitter and hormone releases, muscle contraction, gene expression regulation, and pacemaking by letting calcium ion ($Ca^{2+}$) in many different cells. . The $Ca^{2+}$ that comes in functions as a second messenger and controls such diverse molecular processes. Therefore, dysfunction of Cav channels often cause serious biological disorders. Many neurological or neuromuscular diseases have been shown to involve aberrant Cav functions, such as neuropathic pain, epilepsy, congenital migraine, night blindness, autism spectrum disorders, ataxia, hypokalemic periodic paralysis, Lambert-Eaton syndrome, Timothy syndrome, Brugada syndrome, *etc.* [175]. The Cav channels are protein complexes consisting of a pore forming α1 subunit and auxiliary α2δ, β, and γ subunits. α1 subunit conducts $Ca^{2+}$ upon membrane voltage changes, while auxiliary subunits carry out channel protein trafficking and gating modulation. The α1 subunits are encoded by 10 genes. They are classified into L-type (Cav1.1-1.4), P/Q-type (Cav2.1), N-type (Cav2.2), R-type (Cav2.3), and T-type (Cav3.1-3.3) [176], which have different voltage dependence, biophysical property, and pharmacology. L-, P/Q-, N-, and R-type Cav channels, are among high voltage-activated channels, according to classical terms. Their voltages for half-maximal activation are around -20 to 0 mV. These channels can be further divided by their distinct pharmacological sensitivities. L-type Cavs are blocked by small organic compounds such as dihydropyridines, phenylalkylamines, and benzothiazepines. P/Q-, N-, and R-type Cav channels are blocked by peptide neurotoxins such as ω-agatoxin IVA, ω-conotoxin GVIA, and SNX-482, respectively. Low voltage-activated channels, T-type channels, are activated with half voltage activation of -44 to -46 mV. The channels start to open around the resting membrane potential. Therefore, its function is important for pacemaking and repetitive firing. They have fast inactivation kinetics like Nav channels. Unfortunately, there is no selective blocker for T-type Cav channels. Kurtoxin from scorpion inhibits Cav3.1 with high affinity of 15 nM. Spider toxins ProTx-I and –II also block these channels. But, all those toxins differentially act on other voltage-gated channels. As for small organic compounds, mibefradil shows fair selectivity for T-type over L-type Cav channels.

Calcium channels blockers are well known to reduce cardiac and vascular smooth muscle contractility by blocking L-type Cav channels of the tissues. Many dihydropyridines (for example, nifedipine), phenylalkylamines (for example, verapamil), and benzothiazepines (for example, diltiazem) are being prescribed for hypertension, angina, arrhythmia and other various purposes to control cardiovascular functions [3]. These are considered to be ultimately beneficial against neurological disorders secondarily caused by brain hemorrhage, stroke, *etc*. Other types of Cav channels display relatively high expression profiles in the nervous system. In fact, when discovered, some names of channels reflected the expresser tissues: N-type originated from 'neuronal' and P-type from cerebellar 'Purkinje' neurons (cf. others followed properties: R-type is 'residual' and T-type is 'transient'). Thus, main target tissues will be likely among ones of CNS for future agents modulating the Cav channels. Otherwise, even if a target is outside CNS, the neural tissues should still be seriously treated as a potential generator of adverse effects. Thus, the presence or absence and interpretation of CNS effects should be carefully handled all the time when Cav channels are studied at a therapeutic aspect.

One of good examples above and also for toxin development is N-type Cav channel. N-type Cav channel is emerging as a painkilling target [177]. N-type Cav channels are highly expressed in the superficial lamina (I and II) of the spinal cord, where pain-specific synapses are densely populated between central axons of nociceptor sensory neurons and dendrites of secondary afferents on the pain nerve pathway. In presynapses, N-type Cav channels play a central role in releasing excitatory or peptidergic pain neurotransmitters including glutamate and substance P [178]. Drugs targeting this channel and neighboring signal cascades have long been thought to be potential painkillers, such as N-type Cav channel blockers, NMDA receptor antagonists, neurokinin receptor antagonists, *etc*.

Some of N-type Cav channel blockers were successfully developed and are now clinically being prescribed for pain control. For example, gabapentin, pregabalin, and, ziconotide are available for chronic pain such as neuropathic pain and cancer pain. Interestingly, ziconotide is a synthetic version of a native peptide neurotoxin from Cone snails. The development of this drug from peptide neurotoxins raises possibilities for therapeutic developments using other venom peptides from toxic animals. The primary structure of α1 subunits of Cav channels are highly related to those of Nav channels. The proteins have four homologous domains that are equipped in each with six transmembrane segments.

## Peptide Neurotoxins for Cav Channels

Compared to peptide neurotoxins for Kv channels and Nav channels, there are less number of peptide neurotoxins for Cav channels. In spite of the underdeveloped status, several toxins are displaying promising performance in basic pharmacological studies or drug developments and it is likely because of its high affinity and selectivity.

### *Pore blockers*

ω-conotoxin GVIA, a 27 amino acid toxin, has been isolated from *Conus geographus*. The toxin irreversibly ablates $Ca^{2+}$ components of action potentials and was shown to block N-type Cav channels [179-182]. Another ω-conotoxin MVIIA (ziconotide), a 25 amino acid toxin, has been isolated from *Conus magus*. The toxin is also determined to be a pore blocker and it also acts better on N-type Cav channels [183]. Since then, other several ω-conotoxins and spider toxins acting on Cav channels have been reported (Table **4**) [184-188]. It has been shown that the binding sites of ω-conotoxins are located in S5-S6 region of domain III, according to experiments on P/Q-type and N-type chimeric channels, indicating that ω-conotoxins act as a pore blocker [189]. The binding site turned out to be a putative EF-hand motif in the S5-S6 region of domain III and surprisingly, the external EF-hand motif also affects the relative barium:calcium permeability, which suggests that this toxin competes with $Ca^{2+}$ ion [190]. Interestingly, ω-conotoxin GVIA has been pseudo-irreversible (irreversible but in a typical non-covalent binding manner, not by covalent modifications) and ω-conotoxin MVIIA also less reversible on N-type Cav channels. Stoker *et al.* showed that strong hyperpolarization (-120 mV) renders the toxin binding reversible, suggesting that toxins have binding preference toward inactivated state and thus slowly dissociate [191]. Feng *et al.* suggested a different angle. They showed that the mutation of Gly1326Pro in the putative EF-hand motif made the bindings of both GVIA and MVIIA reversible, suggesting that the reversibility more depends on toxin-channel interaction itself than gating states [192]. The same group further demonstrated that the toxin association is neither voltage-dependent nor frequency-dependent due to their pore occluding mechanism probably by docking the outer vestibule [193]. Surprisingly, it has been recently suggested that the ω-conotoxin GVIA induced gating modification is likely due to allosteric effects of pore binding or electrostatic effects of the positively charged toxin [194, 195]. The two conotoxins MVIIA and GVIA have not only contributed to understanding of physiological functions of N-type Cav channel and its target validation for an analgesic purpose, but also to defining subunit composition [196].

**Table 4:** Peptide neurotoxins for Cav channels

| Toxin name | Source | Sequence | Activity | Potential effect |
|---|---|---|---|---|
| ω-conotoxin GVIA | *Conus geographus* | C KSPGSS C SPTSYN CC RS C NPYTKR C Y* | Cav2.2 inhibition | Ischemic brain injury and pain |
| ω-conotoxin MVIIA | *Conus magus* | C KGKGAK C SRLMYD CC TGS C RSGK C* | Cav2.2 inhibition | Ischemic brain injury and pain |
| ω-conotoxin MVIIC | *Conus magus* | C KGKGAP C RKTMYD CC SGS C GRRGK C* | Cav2.1 inhibition | Chronic pain |
| ω-conotoxin CVID | *Conus cactus* | C KSKGAK C SKLMYD CC TGS C SGTVGR C* | Cav2.2 inhibition | Chronic pain |
| ω-conotoxin CVIE | *Conus cactus* | C KGKGAS C RRTSYD CC TGS C RSGR C* | Cav2.2 inhibition | Neuropathic pain |
| ω-conotoxin CVIF | *Conus cactus* | C KGKGAS C RRTSYD CC TGS C RLGR C* | Cav2.2 inhibition | |
| ω-conotoxin FVIA | *Conus fulmen* | C KGTGKS C SRIAYN CC TGS C RSGK C* | Cav2.2 inhibition | Neuropathic pain |
| DW13.3 | *Filistata hibernalis* | AE C LMIGDTS C VPRLGRR CC YGAW C Y C DQQLS C RRVGRK RE C GWVEVN C K C GWSWSQRIDDWRADYS C K C PEDQ | Cav2.1 inhibition | Pain |
| ω-agatoxin IVA | *Agelenopsis aperta* | KKK C IAKDYGR C KWGGTP CC RGRG C I C SIMGTN C E C KPRLIMEGLGLA | Cav2.1 inhibition | Neuropathic pain |
| ω-grammotoxin SIA | *Grammostola spatulata* | D C VRFWGK C SQTSD CC PHLA C KSKWPRNI C VWDGSV | Cav2.1 inhibition | Pain (predicted) |
| ω-theraphotoxin Hg1a (SNX-482) | *Hysterocrates gigas* | GVDKAGCRYMFGG C SVNDD CC PRLG C HSLFSY C AWDLTFSD | Cav2.3 inhibition | Neuropathic pain |

*Table 4: contd...*

| ProTx-I | *Thrixopelma pruriens* | E C RYWLGG C SAGQT CC KHLV C SRRHGW C VWDGTFS | Cav3.1 inhibition | Epilepsy (predicted) |
|---|---|---|---|---|
| ProTx-II | *Thrixopelma pruriens* | Y C QKWMWT C DSERK CC EGMV C RLW C KKKLW | Cav3.1 inhibition | Epilepsy (predicted) |
| Kurtoxin | *Parabuthus transvaalicus* | KIDGYPVDYWN C KRI C WYNNKY C NDL C KGLKADSGY C WGWTLS C Y C QGLPDNARIKRSGR C RA | Cav3.1, Cav3.2 inhibition | Epilepsy (predicted) |
| PnTx3-6 | *Phoneutria nigriventer* | A C IPRGEI C TDD C ECCG C DNQ C Y C PPGSSLGIFK C S C AHANKYF C NRKKEK C KKA | Cav2.1 inhibition | Chronic pain |
| Huwentoxin-I | *Ornithoctonus huwena* | A C KGVFDA C TPGKNE CC PNRV C SDKHKW C KWKL | Cav2.2 inhibition | Pain in rheumatoid arthritis |
| Huwentoxin-X | *Ornithoctonus huwena* | K C LPPGKP C YGATQKIP CC GV C SHNK C T | Cav2.2 inhibition | Pain (predicted) |

Z, pyroglutamate; O, hydroxyproline; *, -NH2; background colors (pale blue, pink, and pale green) represent pore blockers, gating modifiers, and others, respectively.

Since the findings of the two conotoxins, several other ω-conotoxins have been isolated. CVID (AM336) isolated from *Conus cactus* exhibits an improved selectivity for N-type over P/Q-type, which may implicate its therapeutic potentials (see below) [184]. Recently, CVIE and CVIF from *Conus cactus*, and FVIA from *Conus fulmen* have been reported to reversibly act on N-type Cav channels [197]. Other ω-conotoxin MVIIC from *Conus magus* was shown to inhibit ω-conotoxin GVIA-resistant currents, which are known for P/Q-type Cav currents [198]. Later, it has been turned out that the toxin inhibits N-type Cav channels as well as P/Q-type Cav channels [199].

Several spider peptide neurotoxins have been characterized. They also appear to bind to the pore region of Cav channels. For example, a large spider toxin DW13.3 isolated from *Filistata hibernalis* consists of 74 amino acids and blocks high voltage-activated Cav channels with selectivity for P/Q-type Cav channels. DW13.3 has been determined as a pore blocker based on a 1:1 binding mode and competes with MVIIC [188]. PnTx3-6 isolated from *Phoneutria nigriventer* blocks high voltage-activated Cav channels with a preference of N-type Cav channels [200]. Huwentoxin-I and huwentoxin-X isolated from *Ornithoctonus huwena* also inhibit N-type Cav channel activity. However, the detailed blocking mechanism of spider toxins should be more explored.

## Gating Modifiers

The representative examples of gating modifiers for Cav channels are ω-agatoxin IVA (P/Q-type), ω-grammotoxin SIA (P/Q- and N-type), SNX-482 (R-type), ProTx-I, ProTx-II, and kurtoxin (T-type) [201-205]. Among these gating modifiers, ProTx-I, ProTx-II, SNX-482, and kurtoxin are also Nav channel gating modifiers, reflecting the similarity of Cav and Nav channels [206]. It is interesting that many of pore blockers have been isolated from Cone snails while gating modifiers from spiders (cf. kurtoxin is from scorpion). The toxins retard the channel activation, and accelerate the channel deactivation. A representative Cav gating modifier ω-agatoxin IVA, a 48 amino acid toxin, has been isolated from *Agelenopsis aperta*. It selectively inhibits P/Q-type Cav channels [207]. Upon the toxin application, the activation curve of Cav channels is shifted to more depolarized voltages. ω-agatoxin IVA is known to bind to Glu at the C-terminal end of S3 within domain IV [208]. ω-theraphotoxin Hg1a (SNX-482), a 41 amino acid toxin, was isolated from *Hysterocrates gigas* [202, 209]. The name theraphotoxin is not related to its therapeutic potential but originates from the genus name (Theraphosa) of the source spider. Domains III and IV of R-type Cav channels are binding sites for its gating modification [202, 209]. The toxin also blocks other types of Cav channels and Nav channels [210]. ω-grammotoxin SIA, isolated from *Grammostola spatulata,* is a 36 amino acid toxin blocking N-type and P/Q-type Cav channels. The toxin shifts the channel activation curve to more depolarized voltages like ω-agatoxin IVA but their binding sites are distinct [211, 212]. Kurtoxin from *Parabuthus transvaalicus* has been reported as the first peptide neurotoxin selective for Cav3.1 [203]. However, it was turned out to block high voltage-activated channels as well as Cav3.1 and Cav3.2 [206]. ProTx-I and ProTx-II can block Nav channels and Kv2.1 by gating modification and seem to have a therapeutic potential regarding their Nav1.7 blocking effect with a picomolar $IC_{50}$ (see above). ProTx-I also acts on Cav3.1 with nanomolar affinity [153]. Later, both ProTx-I and ProTx-II have been shown to interact with S3-S4 linker of the domain IV of human Cav 3.1 [204, 205].

## Molecular Properties of Peptide Neurotoxins Targeting Cav Channels

Pore blockers characterized from cone snails employ ICK motif and consist of short three anti-parallel β-sheets and several turn reversals. The structures of these toxins have positive charges by net, like other pore blockers for Kv and Nav channels. The positively charged residues play an important role in binding to the pore of Cav channels along with hydrophobic Tyr residue. For example, alanine scanning mutagenesis studies demonstrated that Lys2 and Tyr13 of ω-conotoxin MVIIA and GVIA are critical residues for blocking N-type Cav channels (Fig. **5**)

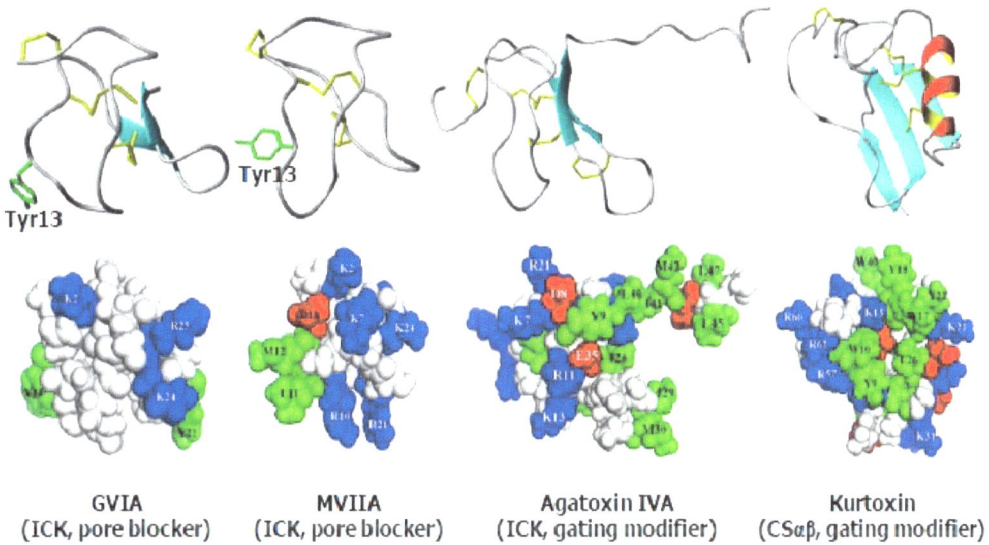

| GVIA | MVIIA | Agatoxin IVA | Kurtoxin |
| (ICK, pore blocker) | (ICK, pore blocker) | (ICK, gating modifier) | (CSαβ, gating modifier) |

**Figure 5:** Three dimensional structures of peptide neurotoxins for Cav channels. Ribbon structures and space filling models of pore blockers GVIA and MVIIA, and gating modifiers agatoxin IVA and kurtoxin. The most important residue Tyr13 of GVIA and MVIIA are colored by green. The disulfide bonds are colored by yellow. In space filling models, hydrophobic, basic, and acidic residues are colored green, blue, and red, respectively. The structures of GVIA, MVIIA, agatoxin IVA, and kurtoxin were obtained from PDB entries 2CCO, 1OMG, 1OAV, and 1T1T, respectively.

[213-215]. Even removal of hydroxyl group by substitution of Tyr13Phe in ω-conotoxin GVIA impaired its blocking activity [216]. Other conotoxins targeting N-type Cav channels (CVID, SO-3, and FVIA) also contain these two residues, reminiscing of the functional dyad in Kv channel blockers. Given that active residues (*e.g.,* Lys2, Tyr13, and Arg21 of ω-conotoxin MVIIA) are located in same surface of inter-cysteine loop2 and loop4, it is likely that residues located in inter-cysteine loop2 also play a role for reversibility. The substitutions of Arg10

in ω-conotoxin MVIIA or Hyp10 in ω-conotoxin CVID with Lys10 have increased their reversibility [217]. Surprisingly, NMR spectrum analysis on MVIIA demonstrated that the inter-cysteine loop2 shows their slow conformational exchange [218]. Other group suggested that the well-defined structure of inter-cysteine loop2 might give a better selectivity to ω-conotoxin CVID for N-type over P/Q-type Cav channels [219]. Accordingly, it is likely that there is a spectrum in conotoxins, from single conformers with a more rigid structure to a slow exchangers and rapid exchangers with a flexible loop structure. In particular for more flexible conformational exchangers, it might be more difficult to chemically follow their scaffolds when one designs toxin-like small molecule drug candidates with a rigid backbone. Probably, detailed information on the fine architecture dependent on aqueous medium (also even on crystalline) is helpful to construct optimal backbones that maximize the drug binding and resultant functionality and to translate the causality for reproductions and applications to other toxins. This appears to be a pivotal step that efforts and advances in technology of structural and pharmaceutical chemistry contribute to boosting up the developments of toxin-mimicking therapeutics in the near future. Because of such importance of inter-cysteine loop2 having β-turn structure, multiple groups are trying to highlight the structure in their developing small molecule inhibitors [220, 221].

Except for kurtoxin, a T-type Cav channel gating modifier, all gating modifiers (ω-agatoxin IVA, ω-grammotoxin SIA, SNX-482, ProTx-I, and ProTx-II) adopt ICK motif in which the third disulfide bond passes through a ring of the first two disulfide bonds [222, 223]. Different from ω-conotoxin MVIIA and GVIA, the structure-function relationship has not been intensively studied because syntheses of these hydrophobic toxins are technically challenging. Nevertheless, it is likely that the residues in hydrophobic patch in their structures play an important role according to the sequence similarity and cross-reactivity between gating modifiers of Kv channels and Cav channels [201]. Kurtoxin, a gating modifier of Cav3.1 and 3.2, is highly homologous with scorpion α-toxins targeting Nav channels. It has a CSαβ motif structure stabilized by four disulfide bonds [224]. Their detailed backbone structural comparison showed that the region in first long loop (Asp8-Ile15) and C-terminal (Arg57-Ala63) have big differences from those of a scorpion α-toxin AaH II. The differences might affect their functional differences but it should be more discussed in the future.

## Therapeutic Implications

N-type Cav channels are found in many regions of CNS including cortex, cerebellum, hippocampus but its function in the spinal cord is most extensively studied. Being uniquely expressed in lamina I and II synapses of spinal cords, N-type Cav plays an important role in pain signaling (see above). Use of toxins already contributed to observing physiology of the channel since selective small molecule modulators were yet to occur then. Even affinity purification of the channel has been carried out using toxins (see review [225]). The function of N-type Cav channels in pain mediation has also been characterized by ω-conotoxins GVIA and MVIIA. Originally, GVIA and MVIIA had been considered as a neuroprotective agent against ischemic brain injury [226-228]. However, accumulating evidence showed that ω-conotoxins GVIA and MVIIA behave as effective analgesic agents by the similar mechanism but in different locations, simultaneously verifying that N-type Cav channels in those locations (the spinal synapses) are an important target in pain signaling [229, 230].

Consistent results were produced by knockout studies. Considering of its wide expression throughout CNS, the phenotypes were somewhat striking. Three independent knockout strains were generated in the same year (2001) and they commonly reported abnormal pain phenotypes despite some technical discrepancies [231-234] . Pathological pain (neuropathic pain and inflammatory pain) was obviously affected. On the other hand, all other vital signs and behaviors are in a normal range but anxiety-associated behaviors. Thus, the genetic approaches firmly supported the hypothesis from toxin pharmacology.

P/Q-type Cav channels also appear to be important in synaptic transmission in the spinal lamina I and II (see below). However, N-type Cavs have differential expression patterns compared to those of P/Q types. When peptidergic neurotransmitter colocalizations were tested, more superimposing features were detected with N-type Cav channels [235, 236]. Peptidergic transmitters (substance P and calcitonin gene-related peptide) and their expressing nociceptors are known to play critical roles in maintaining chronic pain. Therefore, N-type Cav channels might be a major player in the chronic states.

GVIA is a founding member of ω-conotoxins. The clinical developments of the GVIA was hampered by its pseudo-irreversibility (see above) [237]. The irreversibility indicates the difficulty in controlling duration, which leads to narrow therapeutic windows. In this regard, the less irreversibility of MVIIA was advantageous. MVIIA has shown powerful analgesic effects on both acute pain and chronic pain [177, 238]. MVIIA showed analgesic effects on refractory pain caused by AIDS [239]. This gave us two important hints, of which one is biomedical and the other is clinical: N-type Cav channels are an important target for chronic pain [240]. This suggestion can be supported by the fact that upregulation of N-type Cav channels occurred when neuropathic injuries were given [241-243]. Clinically, it is important that toxin treatments turned out to be promising for chronic pain control. MVIIA (ziconotide, SNX-111, or Prialt® from Elan corp. and Azur Pharma) eventually became the first FDA-approved peptide neurotoxin (Table **5**). In practice, ziconotide is supplemented by L-methionine which prevents oxidation of Met12 , its subsequent destabilization and loss of activity. It is intrathecally administered using a catheter and pump infusion system to target N-type Cavs in the spinal cord since it does not readily cross the blood brain barrier. Unexpectedly, despite this being the first case of toxin drug, the length of time taken for the drug development was comparable to those for small molecules (although it started to be purified in early 1980s, its antinociceptive effects were found in early 1990s, and then it was approved by US FDA in 2004 (two month later also by European Medicines Agency)) [177, 238]. It may be because the great advantages of ziconotide are high efficacy for chronic pain and no tolerance [244]. Major problems of currently available analgesics are low efficacy (of nonsteroidal antiinflammatories: NSAIDs), severe tolerance and side effects (of opioids). About a half of total patients suffering from chronic pain are experiencing the efficacy problems. Opioids produce various side effects such as nausea, vomiting, constipation, itching, *etc.* NSAIDs also have gastric problems and thromboembolism-related cardiovascular effects. Ziconotide does not have tolerance [238]. However, some neurological and cardiovascular side effects (representatively, postural hypotension) have been reported. It has been suggested that the adverse effects result from its low selectivity for channels at higher doses or from the N-type Cav blockade in autonomic nerves [245, 246]. Therefore, ziconotide is managed by using a low initial dose and then slow titration in clinics [247]. In these regards, several groups try to improve its therapeutic index. Some groups also try to improve the uncomfortable intrathecal administration way.

**Table 5:** Peptide neurotoxins in clinics and under development

| Peptide Neurotoxins | Sources | Diseases | Targets | Status | Companies |
|---|---|---|---|---|---|
| κ-PVIIA (CGX-1051) | *Conus purpurascens* | Cardiac infarction | Kv heterotetramer | Preclinical | Cognetix Inc. |
| ShK-192 | *Stichodactyla helianthus* | Multiple sclerosis | Kv1.3 | Phase I | Airmid Inc. and Kineta, Inc. |
| MrVIB (CGX-1002) | *Conus marmoreus* | Chronic pain | Nav1.8 | Preclinical | Cognetix Inc. |
| MVIIA (SNX-111 or Ziconitide) | *Conus magus* | Chronic pain | Cav2.2 | US FDA approved in 2004 | Azur Pharma |
| CVID (CNSB004 or Leconotide) | *Conus catus* | Chronic pain | Cav2.2 | Phase I | Relevare Pharmaceuticals |

Successful entry into market of ziconotide is stimulating research to find a therapeutic candidate acting on N-type Cav channels. Its remarkably potent efficacy enough to cover some of opioid-tolerant patients is promoting those, too. At least four hopes were realized by ziconotide development: ziconotide is the first toxin that is approved. Ziconotide is the first marine product that is approved. This is a rare case that an ion channel modulator shows a comparable analgesic potency with opioids with less adverse effects. Teleologically, N-type Cav channels are confirmed to be a valid target to control chronic pain. Therefore, the ziconotide case opens a large number of opportunities to multiple biomedical fields. Even small molecule developments are also becoming active, targeting N-type Cav channels [248]. Together, basic information on the molecular characteristics of N-type Cav channels is growing. N-type Cavs are found both in CNS and PNS (PNS is the primary target for the analgesic effect of ziconotide although it's infused centrally because N-type Cavs function in presynaptic terminals). Interestingly, different splicing variants appear to be expressed in different regions [7, 8, 249, 250]. Sensory neuronal specific variants are reported. Such variation might offer specificity to a certain subtype in a limited brain region of a neurotoxin or a small molecule modulator, which may raise the possibility to identify pain-specific drugs and other CNS-disease specific drugs. Even sensory modality would be selectively controlled if exon37a and exon37b isoforms are discriminated because exon 37a isoform is connected with thermal and mechanical hyperalgesia under inflammation and neuropathy whereas both exon37a and exon37b participate in neuropathic mechanical allodynia [250]. Involvement in G-protein-coupled receptor signaling and detailed molecular processes of N-type-evoked neurotransmitter releases provide mechanisms of the analgesic action of N-type Cav blockers and further validate it as painkilling target [251, 252]

ω-conotoxin CVID from *Conus catus* (AM336, CNSB004 or Leconotide, from Relevare Pharmaceuticals) has exhibited P/Q-type selectivity *in vitro* and low toxicity *in vivo* [240, 253]. Intravenously injected CVID has shown analgesic effects on pathologic pain of diabetes and cancer pain models [253, 254], indicating that a more comfortable administration is possible. This toxin is being developed as pain therapeutics and is currently in clinical trials phase I [255]. CVID tends to be 10-fold weaker in inhibition of vascular sympathetic nerve responses, which may account for less postural hypotension by CVID than those due to MVIIA and GVIA after intravenous administration [256]. Moreover, 2 mg/kg CVID did not cause cardiovascular effects while the non-toxic dose of MVIIA should be as low as 20 μg/kg [253]. CVID (20 μg/kg) showed 25.3 ± 7.6% reversal of hyperalgesic effects and 84.1 ± 7.2% with combination of flupirtine, a Kv channel modulator [253]. MVIIA is more hydrophilic because of charged residues, so that the intravenous toxins can unlikely access target sites across the blood brain barrier. CVID might have better accessibility and also possibly use an unknown novel mechanism to relieve pain as well as N-type Cav channel blocking.

In fact, considering several channelopathic disorders related with P/Q-type mutations are already known, it is a little surprising that potential indications of toxins are limited to pain control until now (see review [257]). Familial hemiplegic migraine type 1 (FHM1), episodic ataxia type 2 (EA2), and spinocerebellar ataxia type 6 (SCA6) are examples. Dozens of mutation sites have been identified for those diseases. For FHM1 and SCA6, gain of function of channels (for example, lowering voltage-threshold for activation) seems to be predominant according to heterologous expression data, but the data should be carefully confirmed since non-neuronal heterologous overexpression might sometimes give a deviated *in vitro* channel behavior. Many of the mutation sites are located in pore region and voltage sensor domain, which reversely emphasizes the functional importance of these two regions. This concept is similarly working for toxin categorization with which toxins are generally classified into pore blockers and gating modifiers as shown in this chapter. Loss of function or even unaltered function was also observed and not only biophysical properties but also expression level and fate and splice variations should not be out of view. EA2 is usually due to loss of function but from mutation to mutation the causes are also not common: biophysical dysfunction of channels, translational problems, dominant negative effects, impaired trafficking, *etc.*

Migraine is a form of pain and migraine and chronic pain largely share a common mechanism: neurogenic inflammation [117]. That is, communication between trigeminal nociceptors and meningeal vascular systems using mediators like calcitonin gene-related peptide and cytokines, amplify the afferent signals, which causes plastic changes in central synapses, in this case, of trigeminal nucleus caudalis or its upper levels. Cortical spreading depression (CSD: slow propagation of depolarization that is experimentally initiated at a focal region of cortex) appears as another mechanism, particularly well accounting for migraine with aura. CSD is potentiated in transgenic mutant mice with FHM1 mutations ([258]; see review [259]). Probably due to cerebellar roles of P/Q-type Cav channels, symptoms such as slowly progressive cerebellar ataxia or nystagmus with cerebellar atrophy accompany FHM1 in cases of certain mutations. EA2 and SCA2 symptoms are fundamentally based on the regional roles of the channel for contributing to coordination of movement and maintenance of balance by cerebellum.

Like N-type Cavs, P/Q-type channels play an important role in initiating neurotransmitter release upon action potential arrival at presynapses. In the cerebellum, P/Q-type Cav channels are responsible at both excitatory and inhibitory synapses (see review [257]). For excitatory ones, the channels work for parallel fibers onto Purkinje neurons, or onto inhibitory interneurons, and climbing fibers onto Purkinje neurons, or onto deep cerebellar nuclei neurons. For inhibitory connections, molecular layer interneurons (basket cells, stellate cells) onto Purkinje neurons, Purkinje neurons onto granule cells, onto deep cerebellar nuclei neurons, or onto other Purkinje neurons. In addition, coupled to the $Ca^{2+}$-activated $K^+$ channels, P/Q-type Cav channels contribute to maintenance of spontaneous intrinsic pacemaking of Purkinje neurons. It is notable that multiple P/Q-type splice variants are in differential charge of the roles in different subregions and subcellular locations. Splice variants exhibit different electrical properties, and even different vulnerability to a gating modifier ω-agatoxin IVA. In cerebral cortical areas, P/Q-type Cav channels seem to take part in synaptic releases at pyramidal neuronal synapses of particular subregions.

ω-conotoxin FVIA from *Conus fulmen* has a better reversibility both in *in vitro* and *in vivo* compared to ω-conotoxin MVIIA [185]. The reversibility is an important factor, given that GVIA failed to be in a clinical setting mainly due to its irreversibility. Phα1β (PnTx3-6) has longer duration of its analgesic actions for postoperative pain and less motor dysfunction [260].

No obvious clinically promising neurotoxin that attacks T-type Cvs occurs yet. However, evidence is increasing that T-type channels are a possible CNS target for pain relief. Due to limited availability of subtype-specific modulators, functional definitions were carried out mostly by *in vivo* or *in vitro* studies using knockout animals or knockdown strategies. In visceral pain, Cv3.1 and Cv3.2 appear to play differential role. Expectedly, Cv3.1 knockouts exhibited reduced visceral pain probably owing to ablated Cv3.1 function in DRG neurons [261]. By contrast, Cv3.2 knockouts showed enhanced visceral pain maybe due to elevated firing of thalamocortical neurons [262]. In inflammatory pain and neuropathic pain, the tendencies were similar [263, 264]. Like N-type Cavs, the differences induced by genetic approaches likely reflect the actions in spinal central synapses. Chemical blockade confirmed the genetic results, which suggests that a toxin modulation may also cause beneficial outcomes [265]. Roles of Cav3.3 remain to be determined.

Some forms of generalized seizures and sleep disorders are also subject to T-type Cav channel dysfunction. Cav3.1 is the predominant T-type channel in thalamocortical neurons. Knockout animal assays yielded an obvious decrease of firing, resistance to drug-induced seizure and spontaneous seizure [266, 267]. Cav3.2 and Cav3.3 seem to be involved in spike wave discharge types with cortical origins but evidence should be more accumulating. In fact, for T-type Cav3.2 channels, mutations were found in patients with absence epilepsies and idiopathic generalized epilepsies [268, 269]. Studies with heterologous overexpression of the relevant mutants of Cav3.2 are suffering from their controversially subtle deviations of electrical phenotypes compared to the wild type channel [269-272]. Mutations in Cav3.2 gene were reported in patients with autism spectrum disorder [273]. A group of those patients also have epileptic disorders and the mutant channel displayed decreased electrical responses *in vitro* analysis.

In electroencephalogram during non-rapid eye movement (non-REM) sleep, Cav3.1 generates spindle wave and Cav3.3 generates slow wave, respectively [266, 274, 275]. Genetically Cav3.1-deleted mice displayed reduced total time of sleep mainly due to a decrease in non-REM sleep [274, 275]. Those also showed multiple brief awakenings leading to fragmentation of sleep. The results implicate that T-type Cav channels play a role in sleep and it can be hypothesized activation of the channels may contribute to prolonged sleep. However, debatable results were obtained from pharmacological modulation. Antagonism of T-type Cavs enhanced sleep and reduced the fragmentation, which is contradictory to knockout results while another study suggested that other Cavs would be more important [276, 277]. Further mechanistic studies are required to conclude the role with more specific blockers for T-type Cav channels. Interestingly, with growing evidence that obesity may employ aberrant circadian rhythms, T-type antagonism and related circuits start drawing attention in the context [278]. As such motivity is increasingly supplied by genetical phenotype analyses and pharmacological simulations, niches for toxin research of T-type Cavs both for purpose of potential therapy and of offering pharmacophores by defining its structural interaction are being enlarged. A possibility is also raised in fields out of neural diseases, such as cardiovascular and cancer biology. Expectedly, current small molecule drug candidates acting on T-type Cavs are mostly aiming at chronic pain and sleep disorders, which attract keen interest (see review [279]).

## OUTLOOK

It is often misunderstood that combinatorial chemistry combined with high throughput assay systems with modern technologies have already overtaken natural products in producing therapeutic agents, but it does not seem to be true. Drugs occupy in between the two groups, and furthermore, industry has currently been suffering from high failure rate in small molecule development [280]. Certainly, there is a lesson, for example, for translational approaches, from nature regarding that chemical pools created by ecological pressure may have an advantage in purposively finding a tool. These days, for neurotoxins, we are now realizing their wealth and become more and more elegant in utilization. First of all, resources for peptide neurotoxins are tremendous. Take an example of cone snails. There are ~700 species of cone snails worldwide. Each species of cone snails have their individual venom mixtures, which contain a myriad of peptide neurotoxins (100-200) without overlap. Therefore, only for cone snails, it can be estimated that there are over 70,000 peptide neurotoxin library [281]. Furthermore, post-translational modifications are frequently detected in conotoxins, causing its complexity to multiply [282]. Possani *et al.* also suggested that ~100,000 different peptides exist in scorpion species [283]. The calculations can be applied to spiders, anemones and so on. In late 1970s and early 1980s, peptide neurotoxins were purified and isolated by biochemical methods which include material separations using columns and Edman degradations. Those methods require a huge amount of venom organ samples from animals. Then, the degenerate sequence has been used, which is to clone the toxins reversely. With technological advances in genetics and proteomics, expression sequence tag (EST) sequencing, genomic sequencing, and mass spectrometry have been more and more contributing to uncovering primary sequences of novel peptide neurotoxins [284-287]. These works have been performed in individual labs or in consortium (for example, Congo project, http://www.conco.eu/noname.html). As a result, disclosures of sequences of potential peptide neurotoxins are growing. For better management of these resources, the web-based databases have been developed. Conoserver (http://www.conoserver.org/) [288], Arachnoserver (http://www.arachnoserver.org/) [289] and SCORPION2 (http://sdmc.i2r.a-star.edu.sg/scorpion/) [290, 291] have been developed for conotoxins, spider toxins, and scorpion toxins, respectively. The resources from these databases seem to promote the research on peptide neurotoxins, ion channels and its therapeutic developments and their collaborations.

Practically, there are two major hurdles in the toxin-based drug developments: one is syntheses of natural peptides and the other is functional assays for voltage-gated ion channels. Peptide neurotoxins are synthesized mainly by solid phase peptide synthesis (SPPS) and gene expression in *E. coli.* or other organisms. The success of peptide synthesis depends on its sequence, since it is difficult to synthesize long peptides having a large number of hydrophobic residues using SPPS. Likewise, hydrophobic peptides would be toxic in *E. coli*. Even though syntheses of linear peptides were successful, disulfide bond formation could be a next issue to overcome in producing natural forms. These difficulties make peptide neurotoxins highly cost. Recently, various techniques such as uses of microwave heating and pseudoproline units have been applied and they have been successful [292, 293]. The syntheses of long peptides (>50 amino acids) also become possible by using native chemical ligation. As for molecular folding, seleno-cysteine and high pressure are shown to be useful. High throughput screening of refolding will promote the finding of the optimal folding condition for native neurotoxins [294, 295].

The second hurdle is tardy functional studies compared to increasing numbers of primary sequences disclosure. For high throughput screening assay for voltage-gated ion channels, making model cell lines expressing these channels is essential but still rare. Electrophysiology with conventional patch clamp technique for voltage-gated ion channels, is a slow work. To address the problems, 1) high throughput automatic patch systems have been developed and are currently being improved. 2) high throughput calcium influx assays (for measuring activities of Cav channels) using KCl application on model cells have been being developed. These assay systems have been working well in screening with small molecules. It has been shown that Kv2.2 channel expression in N-type Cav channel stable cell lines was useful in finding use-dependent blockers. The 4 mM and 14 mM KCl of basal levels in the cells gave different responsiveness to 140 mM KCl [296, 297] .

Other than those conventional screening ways, some creative approaches using tethered toxins have been suggested. Lynx1 is an endogenous cysteine-rich peptide tethered at cell membrane by GPI anchor [298]. Inspired by the presence of lynx1, Ibanez *et al.* designed tethered peptide neurotoxins, which include α-conotoxins, nicotinic acetylcholine receptor ligands, and ω-conotoxins, Cav channel blockers. When these constructs are expressed in *Xenopus* oocyte, the tethered toxins blocked their target currents [299]. Furthermore, when a tethered toxin MrVIA is expressed in Nav1.8-specific neurons, it reduced pain behavior as well as Nav currents [173]. Also, when agatoxin IVA has been expressed in a model animal fruitfly *(Drosophila melanogaster)*, their circadian networks were interrupted [300]. This suggests that a large scale genetic screening would be feasible using *Drosophila* system. The advantage of using *Drosophila* is a rapid behavioral screening system, tissue specific functional assay using USA-GAL4 strategy, and the eukaryotic intracellular folding system different from the reductive environments of *E. coli* [301]. Therefore, the technique would be used for finding novel peptide neurotoxins in near future and will lower the hurdles for drug development of peptide neurotoxins [302].

Out of focus in this chapter, utility of the resources both of peptide neurotoxins and of screening systems are not limited to voltage-gated channels. Our body has over 400 types of ion channel members. All those play important roles in physiology and need modification when those show aberrant functions, which frequently cause serious health problems.

## CONCLUSIONS

Peptide neurotoxins isolated from venomous animals are mainly targeting voltage-gated ion channels. Rigid but diverse three-dimensional structures by intramolecular disulfide bond formations fit into binding sites, which confers potent and selective modulations of channel subtypes. Basic research on peptide neurotoxins already contributed to the understanding of the structure and function of voltage-gated ion channels. In addition, their specificity and potency have prompted some of neurotoxins to enter into market or clinical trials in CNS field. Regarding growing numbers of undeveloped neurotoxin resources and technological progresses, it is highly promising that the peptide neurotoxin research will create therapeutic benefits in the near future.

## ACKNOWLEDGEMENTS

This work was supported by the National Research Foundation of Korea Fellowship (NRF-2011-357-C00125) for S.L., the National Research Foundation of Korea Grant (2012000540) for S.W.H., and Korea Health technology R&D Project of Ministry of Health & Welfare Grant (A111373) for S.W.H., Republic of Korea.

## CONFLICT OF INTEREST

The authors confirm that this chapter content has no conflict of interest.

## REFERENCES

[1]     Overington JP, Al-Lazikani B, Hopkins AL. How many drug targets are there? Nat Rev Drug Discov 2006; 5(12):993-6.

[2]     Ashcroft FM. Ion channels and disease : channelopathies. San Diego: Academic Press; 2000.

[3]     Elliott WJ, Ram CV. Calcium channel blockers. J Clin Hypertens (Greenwich) 2011; 13(9):687-9.

[4]     Cosford ND, Meinke PT, Stauderman KA, Hess SD. Recent advances in the modulation of voltage-gated ion channels for the treatment of epilepsy. Curr Drug Targets CNS Neurol Disord 2002; 1(1):81-104.

[5]     Yu FH, Yarov-Yarovoy V, Gutman GA, Catterall WA. Overview of molecular relationships in the voltage-gated ion channel superfamily. Pharmacol Rev 2005; 57(4):387-95.

[6]     Yu FH, Catterall WA. The VGL-chanome: a protein superfamily specialized for electrical signaling and ionic homeostasis. Sci STKE 2004; 2004(253):re15.

[7]     Bell TJ, Thaler C, Castiglioni AJ, Helton TD, Lipscombe D. Cell-specific alternative splicing increases calcium channel current density in the pain pathway. Neuron 2004; 41(1):127-38.

[8]     Lipscombe D, Raingo J. Alternative splicing matters: N-type calcium channels in nociceptors. Channels (Austin) 2007; 1(4):225-7.

[9]     Shin HS, Cheong EJ, Choi S, Lee J, Na HS. T-type $Ca^{2+}$ channels as therapeutic targets in the nervous system. Curr Opin Pharmacol 2008; 8(1):33-41.

[10]    Lee J, Shin HS. T-type calcium channels and thalamocortical rhythms in sleep: a perspective from studies of T-type calcium channel knockout mice. CNS Neurol Disord Drug Targets 2007; 6(1):63-9.

[11]    Lewis RJ, Garcia ML. Therapeutic potential of venom peptides. Nat Rev Drug Discov 2003; 2(10):790-802.

[12]    Lewis RJ, Dutertre S, Vetter I, Christie MJ. Conus venom peptide pharmacology. Pharmacol Rev 2012; 64(2):259-98.

[13]   King GF. Venoms as a platform for human drugs: translating toxins into therapeutics. Expert Opin Biol Ther 2011; 11(11):1469-84.

[14]   Mouhat S, Jouirou B, Mosbah A, De Waard M, Sabatier JM. Diversity of folds in animal toxins acting on ion channels. Biochem J 2004; 378(Pt 3):717-26.

[15]   Tombola F, Pathak MM, Isacoff EY. How far will you go to sense voltage? Neuron 2005; 48(5):719-25.

[16]   Lee SY, MacKinnon R. A membrane-access mechanism of ion channel inhibition by voltage sensor toxins from spider venom. Nature 2004; 430(6996):232-5.

[17]   Beraud E, Chandy KG. Therapeutic potential of peptide toxins that target ion channels. Inflamm Allergy Drug Targets 2011; 10(5):322-42.

[18]   Mouhat S, Andreotti N, Jouirou B, Sabatier JM. Animal toxins acting on voltage-gated potassium channels. Curr Pharm Des 2008; 14(24):2503-18.

[19]   Diochot S, Lazdunski M. Sea anemone toxins affecting potassium channels. Prog Mol Subcell Biol 2009; 46:99-122.

[20]   Castaneda O, Harvey AL. Discovery and characterization of cnidarian peptide toxins that affect neuronal potassium ion channels. Toxicon 2009; 54(8):1119-24.

[21]   Tempel BL, Papazian DM, Schwarz TL, Jan YN, Jan LY. Sequence of a probable potassium channel component encoded at Shaker locus of Drosophila. Science 1987; 237(4816):770-5.

[22]   Papazian DM, Schwarz TL, Tempel BL, Jan YN, Jan LY. Cloning of genomic and complementary DNA from Shaker, a putative potassium channel gene from Drosophila. Science 1987; 237(4816):749-53.

[23]   Tempel BL, Jan YN, Jan LY. Cloning of a probable potassium channel gene from mouse brain. Nature 1988; 332(6167):837-9.

[24]   Gutman GA, Chandy KG, Grissmer S, *et al.* International Union of Pharmacology. LIII. Nomenclature and molecular relationships of voltage-gated potassium channels. Pharmacol Rev 2005; 57(4):473-508.

[25]   Wulff H, Castle NA, Pardo LA. Voltage-gated potassium channels as therapeutic targets. Nat Rev Drug Discov 2009; 8(12):982-1001.

[26]   Mullard A. 2011 FDA drug approvals. Nat Rev Drug Discov 2012; 11(2):91-4.

[27]   Jiang Y, Lee A, Chen J, *et al.* X-ray structure of a voltage-dependent K$^+$ channel. Nature 2003; 423(6935):33-41.

[28]   Long SB, Campbell EB, Mackinnon R. Crystal structure of a mammalian voltage-dependent Shaker family K$^+$ channel. Science 2005; 309(5736):897-903.

[29]   Phillips LR, Milescu M, Li-Smerin Y, *et al.* Voltage-sensor activation with a tarantula toxin as cargo. Nature 2005; 436(7052):857-60.

[30]   Garcia ML, Gao Y, McManus OB, Kaczorowski GJ. Potassium channels: from scorpion venoms to high-resolution structure. Toxicon 2001; 39(6):739-48.

[31]   Garcia-Calvo M, Knaus HG, McManus OB, *et al.* Purification and reconstitution of the high-conductance, calcium-activated potassium channel from tracheal smooth muscle. J Biol Chem 1994; 269(1):676-82.

[32]   Swartz KJ. Tarantula toxins interacting with voltage sensors in potassium channels. Toxicon 2007; 49(2):213-30.

[33]   Terlau H, Boccaccio A, Olivera BM, Conti F. The block of Shaker K$^+$ channels by kappa-conotoxin PVIIA is state dependent. J Gen Physiol 1999; 114(1):125-40.

[34]   Wulff H, Zhorov BS. $K^+$ channel modulators for the treatment of neurological disorders and autoimmune diseases. Chem Rev 2008; 108(5):1744-73.

[35]   Carbone E, Wanke E, Prestipino G, Possani LD, Maelicke A. Selective blockage of voltage-dependent $K^+$ channels by a novel scorpion toxin. Nature 1982; 296(5852):90-1.

[36]   MacKinnon R, Miller C. Mutant potassium channels with altered binding of charybdotoxin, a pore-blocking peptide inhibitor. Science 1989; 245(4924):1382-5.

[37]   MacKinnon R, Heginbotham L, Abramson T. Mapping the receptor site for charybdotoxin, a pore-blocking potassium channel inhibitor. Neuron 1990; 5(6):767-71.

[38]   Miller C. The charybdotoxin family of $K^+$ channel-blocking peptides. Neuron 1995; 15(1):5-10.

[39]   Valdivia HH, Smith JS, Martin BM, Coronado R, Possani LD. Charybdotoxin and noxiustoxin, two homologous peptide inhibitors of the $K^+$ ($Ca^{2+}$) channel. FEBS Lett 1988; 226(2):280-4.

[40]   Tytgat J, Chandy KG, Garcia ML, *et al.* A unified nomenclature for short-chain peptides isolated from scorpion venoms: alpha-KTx molecular subfamilies. Trends Pharmacol Sci 1999; 20(11):444-7.

[41]   Goldstein SA, Miller C. Mechanism of charybdotoxin block of a voltage-gated $K^+$ channel. Biophys J 1993; 65(4):1613-9.

[42]   Miller C. Competition for block of a $Ca^{2+}$-activated $K^+$ channel by charybdotoxin and tetraethylammonium. Neuron 1988; 1(10):1003-6.

[43]   MacKinnon R. Determination of the subunit stoichiometry of a voltage-activated potassium channel. Nature 1991; 350(6315):232-5.

[44]   Doyle DA, Morais Cabral J, Pfuetzner RA, *et al.* The structure of the potassium channel: molecular basis of $K^+$ conduction and selectivity. Science 1998; 280(5360):69-77.

[45]   Goldstein SA, Pheasant DJ, Miller C. The charybdotoxin receptor of a Shaker $K^+$ channel: peptide and channel residues mediating molecular recognition. Neuron 1994; 12(6):1377-88.

[46]   Ferber M, Sporning A, Jeserich G, *et al.* A novel conus peptide ligand for $K^+$ channels. J Biol Chem 2003; 278(4):2177-83.

[47]   Terlau H, Shon KJ, Grilley M, *et al.* Strategy for rapid immobilization of prey by a fish-hunting marine snail. Nature 1996; 381(6578):148-51.

[48]   Kim M, Baro DJ, Lanning CC, *et al.* Alternative splicing in the pore-forming region of shaker potassium channels. J Neurosci 1997; 17(21):8213-24.

[49]   Shon KJ, Stocker M, Terlau H, *et al.* kappa-Conotoxin PVIIA is a peptide inhibiting the shaker $K^+$ channel. J Biol Chem 1998; 273(1):33-8.

[50]   Garcia E, Scanlon M, Naranjo D. A marine snail neurotoxin shares with scorpion toxins a convergent mechanism of blockade on the pore of voltage-gated K channels. J Gen Physiol 1999; 114(1):141-57.

[51]   Ferber M, Al-Sabi A, Stocker M, Olivera BM, Terlau H. Identification of a mammalian target of kappaM-conotoxin RIIIK. Toxicon 2004; 43(8):915-21.

[52]   Bayrhuber M, Vijayan V, Ferber M, *et al.* Conkunitzin-S1 is the first member of a new Kunitz-type neurotoxin family. Structural and functional characterization. J Biol Chem 2005; 280(25):23766-70.

[53]   Shiomi K. Novel peptide toxins recently isolated from sea anemones. Toxicon 2009; 54(8):1112-8.

[54]    Castaneda O, Sotolongo V, Amor AM, *et al.* Characterization of a potassium channel toxin from the Caribbean Sea anemone Stichodactyla helianthus. Toxicon 1995; 33(5):603-13.

[55]    Aneiros A, Garcia I, Martinez JR, *et al.* A potassium channel toxin from the secretion of the sea anemone Bunodosoma granulifera. Isolation, amino acid sequence and biological activity. Biochim Biophys Acta 1993; 1157(1):86-92.

[56]    Cotton J, Crest M, Bouet F, *et al.* A potassium-channel toxin from the sea anemone Bunodosoma granulifera, an inhibitor for Kv1 channels. Revision of the amino acid sequence, disulfide-bridge assignment, chemical synthesis, and biological activity. Eur J Biochem 1997; 244(1):192-202.

[57]    Swartz KJ, MacKinnon R. An inhibitor of the Kv2.1 potassium channel isolated from the venom of a Chilean tarantula. Neuron 1995; 15(4):941-9.

[58]    Swartz KJ, MacKinnon R. Mapping the receptor site for hanatoxin, a gating modifier of voltage-dependent K$^+$ channels. Neuron 1997; 18(4):675-82.

[59]    Swartz KJ, MacKinnon R. Hanatoxin modifies the gating of a voltage-dependent K$^+$ channel through multiple binding sites. Neuron 1997; 18(4):665-73.

[60]    Li-Smerin Y, Swartz KJ. Helical structure of the COOH terminus of S3 and its contribution to the gating modifier toxin receptor in voltage-gated ion channels. J Gen Physiol 2001; 117(3):205-18.

[61]    Lee CW, Kim S, Roh SH, *et al.* Solution structure and functional characterization of SGTx1, a modifier of Kv2.1 channel gating. Biochemistry 2004; 43(4):890-7.

[62]    Wang JM, Roh SH, Kim S, *et al.* Molecular surface of tarantula toxins interacting with voltage sensors in K(v) channels. J Gen Physiol 2004; 123(4):455-67.

[63]    Herrington J, Zhou YP, Bugianesi RM, *et al.* Blockers of the delayed-rectifier potassium current in pancreatic beta-cells enhance glucose-dependent insulin secretion. Diabetes 2006; 55(4):1034-42.

[64]    Lee S, Milescu M, Jung HH, *et al.* Solution structure of GxTX-1E, a high-affinity tarantula toxin interacting with voltage sensors in Kv2.1 potassium channels. Biochemistry 2010; 49(25):5134-42.

[65]    Xiao Y, Tang J, Yang Y, *et al.* Jingzhaotoxin-III, a novel spider toxin inhibiting activation of voltage-gated sodium channel in rat cardiac myocytes. J Biol Chem 2004; 279(25):26220-6.

[66]    Yuan C, Yang S, Liao Z, Liang S. Effects and mechanism of Chinese tarantula toxins on the Kv2.1 potassium channels. Biochem Biophys Res Commun 2007; 352(3):799-804.

[67]    Bode F, Sachs F, Franz MR. Tarantula peptide inhibits atrial fibrillation. Nature 2001; 409(6816):35-6.

[68]    Hill K, Schaefer M. TRPA1 is differentially modulated by the amphipathic molecules trinitrophenol and chlorpromazine. J Biol Chem 2007; 282(10):7145-53.

[69]    Diochot S, Schweitz H, Beress L, Lazdunski M. Sea anemone peptides with a specific blocking activity against the fast inactivating potassium channel Kv3.4. J Biol Chem 1998; 273(12):6744-9.

[70]    Liu P, Jo S, Bean BP. Modulation of neuronal sodium channels by the sea anemone peptide BDS-I. J Neurophysiol 2012; 107(11):3155-67.

[71]    Diochot S, Loret E, Bruhn T, Beress L, Lazdunski M. APETx1, a new toxin from the sea anemone Anthopleura elegantissima, blocks voltage-gated human ether-a-go-go-related gene potassium channels. Mol Pharmacol 2003; 64(1):59-69.

[72]    Zhang M, Liu XS, Diochot S, Lazdunski M, Tseng GN. APETx1 from sea anemone Anthopleura elegantissima is a gating modifier peptide toxin of the human ether-a-go-go-related potassium channel. Mol Pharmacol 2007; 72(2):259-68.

[73]    Teichert RW, Jacobsen R, Terlau H, Yoshikami D, Olivera BM. Discovery and characterization of the short kappaA-conotoxins: a novel subfamily of excitatory conotoxins. Toxicon 2007; 49(3):318-28.

[74]    Massilia GR, Eliseo T, Grolleau F, *et al.* Contryphan-Vn: a modulator of $Ca^{2+}$-dependent $K^+$ channels. Biochem Biophys Res Commun 2003; 303(1):238-46.

[75]    Peng C, Tang S, Pi C, *et al.* Discovery of a novel class of conotoxin from Conus litteratus, lt14a, with a unique cysteine pattern. Peptides 2006; 27(9):2174-81.

[76]    Hu SH, Gehrmann J, Alewood PF, Craik DJ, Martin JL. Crystal structure at 1.1 A resolution of alpha-conotoxin PnIB: comparison with alpha-conotoxins PnIA and GI. Biochemistry 1997; 36(38):11323-30.

[77]    Hu SH, Gehrmann J, Guddat LW, *et al.* The 1.1 A crystal structure of the neuronal acetylcholine receptor antagonist, alpha-conotoxin PnIA from Conus pennaceus. Structure 1996; 4(4):417-23.

[78]    Dy CY, Buczek P, Imperial JS, Bulaj G, Horvath MP. Structure of conkunitzin-S1, a neurotoxin and Kunitz-fold disulfide variant from cone snail. Acta Crystallogr D Biol Crystallogr 2006; 62(Pt 9):980-90.

[79]    Bontems F, Roumestand C, Gilquin B, Menez A, Toma F. Refined structure of charybdotoxin: common motifs in scorpion toxins and insect defensins. Science 1991; 254(5037):1521-3.

[80]    Pallaghy PK, Nielsen KJ, Craik DJ, Norton RS. A common structural motif incorporating a cystine knot and a triple-stranded beta-sheet in toxic and inhibitory polypeptides. Protein Sci 1994; 3(10):1833-9.

[81]    Scanlon MJ, Naranjo D, Thomas L, *et al.* Solution structure and proposed binding mechanism of a novel potassium channel toxin kappa-conotoxin PVIIA. Structure 1997; 5(12):1585-97.

[82]    Dauplais M, Lecoq A, Song J, *et al.* On the convergent evolution of animal toxins. Conservation of a diad of functional residues in potassium channel-blocking toxins with unrelated structures. J Biol Chem 1997; 272(7):4302-9.

[83]    Katoh E, Nishio H, Inui T, *et al.* Structural basis for the biological activity of dendrotoxin-I, a potent potassium channel blocker. Biopolymers 2000; 54(1):44-57.

[84]    Gilquin B, Racape J, Wrisch A, *et al.* Structure of the BgK-Kv1.1 complex based on distance restraints identified by double mutant cycles. Molecular basis for convergent evolution of Kv1 channel blockers. J Biol Chem 2002; 277(40):37406-13.

[85]    Takahashi H, Kim JI, Min HJ, *et al.* Solution structure of hanatoxin1, a gating modifier of voltage-dependent $K^+$ channels: common surface features of gating modifier toxins. J Mol Biol 2000; 297(3):771-80.

[86]    Lubbers NL, Campbell TJ, Polakowski JS, *et al.* Postischemic administration of CGX-1051, a peptide from cone snail venom, reduces infarct size in both rat and dog models of myocardial ischemia and reperfusion. J Cardiovasc Pharmacol 2005; 46(2):141-6.

[87]    Zhang SJ, Yang XM, Liu GS, *et al.* CGX-1051, a peptide from Conus snail venom, attenuates infarction in rabbit hearts when administered at reperfusion. J Cardiovasc Pharmacol 2003; 42(6):764-71.

[88] Chen P, Dendorfer A, Finol-Urdaneta RK, Terlau H, Olivera BM. Biochemical characterization of kappaM-RIIIJ, a Kv1.2 channel blocker: evaluation of cardioprotective effects of kappaM-conotoxins. J Biol Chem 2010; 285(20):14882-9.

[89] Pennington MW, Beeton C, Galea CA, *et al.* Engineering a stable and selective peptide blocker of the Kv1.3 channel in T lymphocytes. Mol Pharmacol 2009; 75(4):762-73.

[90] Beeton C, Wulff H, Barbaria J, *et al.* Selective blockade of T lymphocyte $K^+$ channels ameliorates experimental autoimmune encephalomyelitis, a model for multiple sclerosis. Proc Natl Acad Sci U S A 2001; 98(24):13942-7.

[91] Lu Q, Peevey J, Jow F, *et al.* Disruption of Kv1.1 N-type inactivation by novel small molecule inhibitors (disinactivators). Bioorg Med Chem 2008; 16(6):3067-75.

[92] Smart SL, Lopantsev V, Zhang CL, *et al.* Deletion of the K(V)1.1 potassium channel causes epilepsy in mice. Neuron 1998; 20(4):809-19.

[93] Brew HM, Gittelman JX, Silverstein RS, *et al.* Seizures and reduced life span in mice lacking the potassium channel subunit Kv1.2, but hypoexcitability and enlarged Kv1 currents in auditory neurons. J Neurophysiol 2007; 98(3):1501-25.

[94] Zuberi SM, Eunson LH, Spauschus A, *et al.* A novel mutation in the human voltage-gated potassium channel gene (Kv1.1) associates with episodic ataxia type 1 and sometimes with partial epilepsy. Brain 1999; 122 ( Pt 5):817-25.

[95] Karimi-Abdolrezaee S, Eftekharpour E, Fehlings MG. Temporal and spatial patterns of Kv1.1 and Kv1.2 protein and gene expression in spinal cord white matter after acute and chronic spinal cord injury in rats: implications for axonal pathophysiology after neurotrauma. Eur J Neurosci 2004; 19(3):577-89.

[96] Nashmi R, Fehlings MG. Mechanisms of axonal dysfunction after spinal cord injury: with an emphasis on the role of voltage-gated potassium channels. Brain Res Brain Res Rev 2001; 38(1-2):165-91.

[97] Pannaccione A, Boscia F, Scorziello A, *et al.* Up-regulation and increased activity of KV3.4 channels and their accessory subunit MinK-related peptide 2 induced by amyloid peptide are involved in apoptotic neuronal death. Mol Pharmacol 2007; 72(3):665-73.

[98] Blackburn-Munro G, Dalby-Brown W, Mirza NR, Mikkelsen JD, Blackburn-Munro RE. Retigabine: chemical synthesis to clinical application. CNS Drug Rev 2005; 11(1):1-20.

[99] Munro G, Dalby-Brown W. Kv7 (KCNQ) channel modulators and neuropathic pain. J Med Chem 2007; 50(11):2576-82.

[100] Wickenden AD, McNaughton-Smith G. Kv7 channels as targets for the treatment of pain. Curr Pharm Des 2009; 15(15):1773-98.

[101] Blackburn-Munro G, Jensen BS. The anticonvulsant retigabine attenuates nociceptive behaviours in rat models of persistent and neuropathic pain. Eur J Pharmacol 2003; 460(2-3):109-16.

[102] Hansen HH, Waroux O, Seutin V, *et al.* Kv7 channels: interaction with dopaminergic and serotonergic neurotransmission in the CNS. J Physiol 2008; 586(7):1823-32.

[103] Redrobe JP, Nielsen AN. Effects of neuronal Kv7 potassium channel activators on hyperactivity in a rodent model of mania. Behav Brain Res 2009; 198(2):481-5.

[104] Lehmann-Horn F, Jurkat-Rott K. Voltage-gated ion channels and hereditary disease. Physiol Rev 1999; 79(4):1317-72.

[105] Mantegazza M, Curia G, Biagini G, Ragsdale DS, Avoli M. Voltage-gated sodium channels as therapeutic targets in epilepsy and other neurological disorders. Lancet Neurol 2010; 9(4):413-24.

[106] Catterall WA, Goldin AL, Waxman SG. International Union of Pharmacology. XLVII. Nomenclature and structure-function relationships of voltage-gated sodium channels. Pharmacol Rev 2005; 57(4):397-409.

[107] Goldin AL, Barchi RL, Caldwell JH, *et al.* Nomenclature of voltage-gated sodium channels. Neuron 2000; 28(2):365-8.

[108] Watanabe E, Fujikawa A, Matsunaga H, *et al.* Nav2/NaG channel is involved in control of salt-intake behavior in the CNS. J Neurosci 2000; 20(20):7743-51.

[109] Akopian AN, Sivilotti L, Wood JN. A tetrodotoxin-resistant voltage-gated sodium channel expressed by sensory neurons. Nature 1996; 379(6562):257-62.

[110] Bierhaus A, Fleming T, Stoyanov S, *et al.* Methylglyoxal modification of Nav1.8 facilitates nociceptive neuron firing and causes hyperalgesia in diabetic neuropathy. Nat Med 2012; 18(6):926-33.

[111] Payandeh J, Scheuer T, Zheng N, Catterall WA. The crystal structure of a voltage-gated sodium channel. Nature 2011; 475(7356):353-8.

[112] Payandeh J, Gamal El-Din TM, Scheuer T, Zheng N, Catterall WA. Crystal structure of a voltage-gated sodium channel in two potentially inactivated states. Nature 2012; 486(7401):135-9.

[113] End*o M, Kurachi Y, Mishina M. Pharmacology of ionic channel function : activators and inhibitors. Berlin ; New York: Springer; 2000.

[114] Binshtok AM, Bean BP, Woolf CJ. Inhibition of nociceptors by TRPV1-mediated entry of impermeant sodium channel blockers. Nature 2007; 449(7162):607-10.

[115] Strichartz GR. The inhibition of sodium currents in myelinated nerve by quaternary derivatives of lidocaine. J Gen Physiol 1973; 62(1):37-57.

[116] Hille B. Ion channels of excitable membranes. 3rd ed. Sunderland, Mass.: Sinauer; 2001.

[117] Chiu IM, von Hehn CA, Woolf CJ. Neurogenic inflammation and the peripheral nervous system in host defense and immunopathology. Nat Neurosci 2012; 15(8):1063-7.

[118] Jarvis MF, Honore P, Shieh CC, *et al.* A-803467, a potent and selective Nav1.8 sodium channel blocker, attenuates neuropathic and inflammatory pain in the rat. Proc Natl Acad Sci U S A 2007; 104(20):8520-5.

[119] Tarnawa I, Bolcskei H, Kocsis P. Blockers of voltage-gated sodium channels for the treatment of central nervous system diseases. Recent Pat CNS Drug Discov 2007; 2(1):57-78.

[120] Klint JK, Senff S, Rupasinghe DB, *et al.* Spider-venom peptides that target voltage-gated sodium channels: Pharmacological tools and potential therapeutic leads. Toxicon 2012.

[121] Rodriguez de la Vega RC, Possani LD. Overview of scorpion toxins specific for $Na^+$ channels and related peptides: biodiversity, structure-function relationships and evolution. Toxicon 2005; 46(8):831-44.

[122] Ekberg J, Craik DJ, Adams DJ. Conotoxin modulation of voltage-gated sodium channels. Int J Biochem Cell Biol 2008; 40(11):2363-8.

[123] Cruz LJ, Gray WR, Olivera BM, *et al.* Conus geographus toxins that discriminate between neuronal and muscle sodium channels. J Biol Chem 1985; 260(16):9280-8.

[124] Yanagawa Y, Abe T, Satake M. Mu-conotoxins share a common binding site with tetrodotoxin/saxitoxin on eel electroplax Na channels. J Neurosci 1987; 7(5):1498-502.

[125] Stephan MM, Potts JF, Agnew WS. The microI skeletal muscle sodium channel: mutation E403Q eliminates sensitivity to tetrodotoxin but not to mu-conotoxins GIIIA and GIIIB. J Membr Biol 1994; 137(1):1-8.

[126]  Shon KJ, Olivera BM, Watkins M, *et al.* mu-Conotoxin PIIIA, a new peptide for discriminating among tetrodotoxin-sensitive Na channel subtypes. J Neurosci 1998; 18(12):4473-81.

[127]  Cummins TR, Aglieco F, Dib-Hajj SD. Critical molecular determinants of voltage-gated sodium channel sensitivity to mu-conotoxins GIIIA/B. Mol Pharmacol 2002; 61(5):1192-201.

[128]  Chahine M, Sirois J, Marcotte P, Chen L, Kallen RG. Extrapore residues of the S5-S6 loop of domain 2 of the voltage-gated skeletal muscle sodium channel (rSkM1) contribute to the mu-conotoxin GIIIA binding site. Biophys J 1998; 75(1):236-46.

[129]  Bulaj G, West PJ, Garrett JE, *et al.* Novel conotoxins from Conus striatus and Conus kinoshitai selectively block TTX-resistant sodium channels. Biochemistry 2005; 44(19):7259-65.

[130]  Zhang MM, Gruszczynski P, Walewska A, *et al.* Cooccupancy of the outer vestibule of voltage-gated sodium channels by micro-conotoxin KIIIA and saxitoxin or tetrodotoxin. J Neurophysiol 2010; 104(1):88-97.

[131]  Zhang MM, McArthur JR, Azam L, *et al.* Synergistic and antagonistic interactions between tetrodotoxin and mu-conotoxin in blocking voltage-gated sodium channels. Channels (Austin) 2009; 3(1):32-8.

[132]  Yanagawa Y, Abe T, Satake M, *et al.* A novel sodium channel inhibitor from Conus geographus: purification, structure, and pharmacological properties. Biochemistry 1988; 27(17):6256-62.

[133]  Martin-Moutot N, Mansuelle P, Alcaraz G, *et al.* Phoneutria nigriventer toxin 1: a novel, state-dependent inhibitor of neuronal sodium channels that interacts with micro conotoxin binding sites. Mol Pharmacol 2006; 69(6):1931-7.

[134]  Rogers JC, Qu Y, Tanada TN, Scheuer T, Catterall WA. Molecular determinants of high affinity binding of alpha-scorpion toxin and sea anemone toxin in the S3-S4 extracellular loop in domain IV of the $Na^+$ channel alpha subunit. J Biol Chem 1996; 271(27):15950-62.

[135]  Rochat H, Rochat C, Sampieri F, Miranda F, Lissitzky S. The amino-acid sequence of neurotoxin II of Androctonus australis Hector. Eur J Biochem 1972; 28(3):381-8.

[136]  Chen H, Heinemann SH. Interaction of scorpion alpha-toxins with cardiac sodium channels: binding properties and enhancement of slow inactivation. J Gen Physiol 2001; 117(6):505-18.

[137]  Fainzilber M, Kofman O, Zlotkin E, Gordon D. A new neurotoxin receptor site on sodium channels is identified by a conotoxin that affects sodium channel inactivation in molluscs and acts as an antagonist in rat brain. J Biol Chem 1994; 269(4):2574-80.

[138]  Leipold E, Hansel A, Olivera BM, Terlau H, Heinemann SH. Molecular interaction of delta-conotoxins with voltage-gated sodium channels. FEBS Lett 2005; 579(18):3881-4.

[139]  Barbier J, Lamthanh H, Le Gall F, *et al.* A delta-conotoxin from Conus ermineus venom inhibits inactivation in vertebrate neuronal $Na^+$ channels but not in skeletal and cardiac muscles. J Biol Chem 2004; 279(6):4680-5.

[140]  Cestele S, Qu Y, Rogers JC, *et al.* Voltage sensor-trapping: enhanced activation of sodium channels by beta-scorpion toxin bound to the S3-S4 loop in domain II. Neuron 1998; 21(4):919-31.

[141]  Cohen L, Ilan N, Gur M, *et al.* Design of a specific activator for skeletal muscle sodium channels uncovers channel architecture. J Biol Chem 2007; 282(40):29424-30.

[142] Leipold E, Borges A, Heinemann SH. Scorpion beta-toxin interference with NaV channel voltage sensor gives rise to excitatory and depressant modes. J Gen Physiol 2012; 139(4):305-19.

[143] Leipold E, Hansel A, Borges A, Heinemann SH. Subtype specificity of scorpion beta-toxin Tz1 interaction with voltage-gated sodium channels is determined by the pore loop of domain 3. Mol Pharmacol 2006; 70(1):340-7.

[144] Fainzilber M, van der Schors R, Lodder JC, *et al.* New sodium channel-blocking conotoxins also affect calcium currents in Lymnaea neurons. Biochemistry 1995; 34(16):5364-71.

[145] McIntosh JM, Hasson A, Spira ME, *et al.* A new family of conotoxins that blocks voltage-gated sodium channels. J Biol Chem 1995; 270(28):16796-802.

[146] Daly NL, Ekberg JA, Thomas L, *et al.* Structures of muO-conotoxins from Conus marmoreus. I nhibitors of tetrodotoxin (TTX)-sensitive and TTX-resistant sodium channels in mammalian sensory neurons. J Biol Chem 2004; 279(24):25774-82.

[147] Leipold E, DeBie H, Zorn S, *et al.* muO conotoxins inhibit NaV channels by interfering with their voltage sensors in domain-2. Channels (Austin) 2007; 1(4):253-62.

[148] Terlau H, Stocker M, Shon KJ, McIntosh JM, Olivera BM. MicroO-conotoxin MrVIA inhibits mammalian sodium channels, but not through site I. J Neurophysiol 1996; 76(3):1423-9.

[149] Zorn S, Leipold E, Hansel A, *et al.* The muO-conotoxin MrVIA inhibits voltage-gated sodium channels by associating with domain-3. FEBS Lett 2006; 580(5):1360-4.

[150] Pi C, Liu J, Wang L, *et al.* Soluble expression, purification and functional identification of a disulfide-rich conotoxin derived from Conus litteratus. J Biotechnol 2007; 128(1):184-93.

[151] Wang L, Pi C, Liu J, *et al.* Identification and characterization of a novel O-superfamily conotoxin from Conus litteratus. J Pept Sci 2008; 14(10):1077-83.

[152] Priest BT, Blumenthal KM, Smith JJ, Warren VA, Smith MM. ProTx-I and ProTx-II: gating modifiers of voltage-gated sodium channels. Toxicon 2007; 49(2):194-201.

[153] Middleton RE, Warren VA, Kraus RL, *et al.* Two tarantula peptides inhibit activation of multiple sodium channels. Biochemistry 2002; 41(50):14734-47.

[154] Liu J, Wu Q, Pi C, *et al.* Isolation and characterization of a T-superfamily conotoxin from Conus litteratus with targeting tetrodotoxin-sensitive sodium channels. Peptides 2007; 28(12):2313-9.

[155] Jimenez EC, Shetty RP, Lirazan M, *et al.* Novel excitatory Conus peptides define a new conotoxin superfamily. J Neurochem 2003; 85(3):610-21.

[156] Fiedler B, Zhang MM, Buczek O, *et al.* Specificity, affinity and efficacy of iota-conotoxin RXIA, an agonist of voltage-gated sodium channels Na(V)1.2, 1.6 and 1.7. Biochem Pharmacol 2008; 75(12):2334-44.

[157] Wang L, Liu J, Pi C, *et al.* Identification of a novel M-superfamily conotoxin with the ability to enhance tetrodotoxin sensitive sodium currents. Arch Toxicol 2009; 83(10):925-32.

[158] Hill JM, Alewood PF, Craik DJ. Solution structure of the sodium channel antagonist conotoxin GS: a new molecular caliper for probing sodium channel geometry. Structure 1997; 5(4):571-83.

[159] Housset D, Habersetzer-Rochat C, Astier JP, Fontecilla-Camps JC. Crystal structure of toxin II from the scorpion Androctonus australis Hector refined at 1.3 A resolution. J Mol Biol 1994; 238(1):88-103.

[160]  Pintar A, Possani LD, Delepierre M. Solution structure of toxin 2 from centruroides noxius Hoffmann, a beta-scorpion neurotoxin acting on sodium channels. J Mol Biol 1999; 287(2):359-67.

[161]  Volpon L, Lamthanh H, Barbier J, *et al.* NMR solution structures of delta-conotoxin EVIA from Conus ermineus that selectively acts on vertebrate neuronal $Na^+$ channels. J Biol Chem 2004; 279(20):21356-66.

[162]  Kharrat R, Darbon H, Rochat H, Granier C. Structure/activity relationships of scorpion alpha-toxins. Multiple residues contribute to the interaction with receptors. Eur J Biochem 1989; 181(2):381-90.

[163]  Zimmermann K, Leffler A, Babes A, *et al.* Sensory neuron sodium channel Nav1.8 is essential for pain at low temperatures. Nature 2007; 447(7146):855-8.

[164]  Minett MS, Nassar MA, Clark AK, *et al.* Distinct Nav1.7-dependent pain sensations require different sets of sensory and sympathetic neurons. Nat Commun 2012; 3:791.

[165]  Wood JN, Boorman JP, Okuse K, Baker MD. Voltage-gated sodium channels and pain pathways. J Neurobiol 2004; 61(1):55-71.

[166]  Cox JJ, Reimann F, Nicholas AK, *et al.* An SCN9A channelopathy causes congenital inability to experience pain. Nature 2006; 444(7121):894-8.

[167]  Yang Y, Wang Y, Li S, *et al.* Mutations in SCN9A, encoding a sodium channel alpha subunit, in patients with primary erythermalgia. J Med Genet 2004; 41(3):171-4.

[168]  Fertleman CR, Baker MD, Parker KA, *et al.* SCN9A mutations in paroxysmal extreme pain disorder: allelic variants underlie distinct channel defects and phenotypes. Neuron 2006; 52(5):767-74.

[169]  Drenth JP, Waxman SG. Mutations in sodium-channel gene SCN9A cause a spectrum of human genetic pain disorders. J Clin Invest 2007; 117(12):3603-9.

[170]  Schmalhofer WA, Calhoun J, Burrows R, *et al.* ProTx-II, a selective inhibitor of NaV1.7 sodium channels, blocks action potential propagation in nociceptors. Mol Pharmacol 2008; 74(5):1476-84.

[171]  Saez NJ, Senff S, Jensen JE, *et al.* Spider-venom peptides as therapeutics. Toxins (Basel) 2010; 2(12):2851-71.

[172]  Ekberg J, Jayamanne A, Vaughan CW, *et al.* muO-conotoxin MrVIB selectively blocks Nav1.8 sensory neuron specific sodium channels and chronic pain behavior without motor deficits. Proc Natl Acad Sci U S A 2006; 103(45):17030-5.

[173]  Sturzebecher AS, Hu J, Smith ES, *et al.* An *in vivo* tethered toxin approach for the cell-autonomous inactivation of voltage-gated sodium channel currents in nociceptors. J Physiol 2010; 588(Pt 10):1695-707.

[174]  Zhang MM, Green BR, Catlin P, *et al.* Structure/function characterization of micro-conotoxin KIIIA, an analgesic, nearly irreversible blocker of mammalian neuronal sodium channels. J Biol Chem 2007; 282(42):30699-706.

[175]  Cain SM, Snutch TP. Voltage-gated calcium channels and disease. Biofactors 2011; 37(3):197-205.

[176]  Catterall WA, Perez-Reyes E, Snutch TP, Striessnig J. International Union of Pharmacology. XLVIII. Nomenclature and structure-function relationships of voltage-gated calcium channels. Pharmacol Rev 2005; 57(4):411-25.

[177]  Malmberg AB, Yaksh TL. Voltage-sensitive calcium channels in spinal nociceptive processing: blockade of N- and P-type channels inhibits formalin-induced nociception. J Neurosci 1994; 14(8):4882-90.

[178]  Kerr LM, Filloux F, Olivera BM, Jackson H, Wamsley JK. Autoradiographic localization of calcium channels with [125I]omega-conotoxin in rat brain. Eur J Pharmacol 1988; 146(1):181-3.

[179]  Kerr LM, Yoshikami D. A venom peptide with a novel presynaptic blocking action. Nature 1984; 308(5956):282-4.

[180]  Olivera BM, McIntosh JM, Cruz LJ, Luque FA, Gray WR. Purification and sequence of a presynaptic peptide toxin from Conus geographus venom. Biochemistry 1984; 23(22):5087-90.

[181]  Reynolds IJ, Wagner JA, Snyder SH, *et al.* Brain voltage-sensitive calcium channel subtypes differentiated by omega-conotoxin fraction GVIA. Proc Natl Acad Sci U S A 1986; 83(22):8804-7.

[182]  Olivera BM, Gray WR, Zeikus R, *et al.* Peptide neurotoxins from fish-hunting cone snails. Science 1985; 230(4732):1338-43.

[183]  Olivera BM, Cruz LJ, de Santos V, *et al.* Neuronal calcium channel antagonists. Discrimination between calcium channel subtypes using omega-conotoxin from Conus magus venom. Biochemistry 1987; 26(8):2086-90.

[184]  Lewis RJ, Nielsen KJ, Craik DJ, *et al.* Novel omega-conotoxins from Conus catus discriminate among neuronal calcium channel subtypes. J Biol Chem 2000; 275(45):35335-44.

[185]  Lee S, Kim Y, Back SK, *et al.* Analgesic effect of highly reversible omega-conotoxin FVIA on N type $Ca^{2+}$ channels. Mol Pain 2010; 6:97.

[186]  Wen L, Yang S, Qiao H, *et al.* SO-3, a new O-superfamily conopeptide derived from Conus striatus, selectively inhibits N-type calcium currents in cultured hippocampal neurons. Br J Pharmacol 2005; 145(6):728-39.

[187]  Souza AH, Ferreira J, Cordeiro Mdo N, *et al.* Analgesic effect in rodents of native and recombinant Ph alpha 1beta toxin, a high-voltage-activated calcium channel blocker isolated from armed spider venom. Pain 2008; 140(1):115-26.

[188]  Sutton KG, Siok C, Stea A, *et al.* Inhibition of neuronal calcium channels by a novel peptide spider toxin, DW13.3. Mol Pharmacol 1998; 54(2):407-18.

[189]  Ellinor PT, Zhang JF, Horne WA, Tsien RW. Structural determinants of the blockade of N-type calcium channels by a peptide neurotoxin. Nature 1994; 372(6503):272-5.

[190]  Feng ZP, Hamid J, Doering C, *et al.* Amino acid residues outside of the pore region contribute to N-type calcium channel permeation. J Biol Chem 2001; 276(8):5726-30.

[191]  Stocker JW, Nadasdi L, Aldrich RW, Tsien RW. Preferential interaction of omega-conotoxins with inactivated N-type $Ca^{2+}$ channels. J Neurosci 1997; 17(9):3002-13.

[192]  Feng ZP, Hamid J, Doering C, *et al.* Residue Gly1326 of the N-type calcium channel alpha 1B subunit controls reversibility of omega-conotoxin GVIA and MVIIA block. J Biol Chem 2001; 276(19):15728-35.

[193]  Feng ZP, Doering CJ, Winkfein RJ, *et al.* Determinants of inhibition of transiently expressed voltage-gated calcium channels by omega-conotoxins GVIA and MVIIA. J Biol Chem 2003; 278(22):20171-8.

[194]  Yarotskyy V, Elmslie KS. omega-conotoxin GVIA alters gating charge movement of N-type (CaV2.2) calcium channels. J Neurophysiol 2009; 101(1):332-40.

[195]  Yarotskyy V, Elmslie KS. Interference between two modulators of N-type (CaV2.2) calcium channel gating demonstrates that omega-conotoxin GVIA disrupts open state gating. Biochim Biophys Acta 2010; 1798(9):1821-8.

[196]  Witcher DR, De Waard M, Sakamoto J, *et al.* Subunit identification and reconstitution of the N-type $Ca^{2+}$ channel complex purified from brain. Science 1993; 261(5120):486-9.

[197]  Berecki G, Motin L, Haythornthwaite A, *et al.* Analgesic (omega)-conotoxins CVIE and CVIF selectively and voltage-dependently block recombinant and native N-type calcium channels. Mol Pharmacol 2010; 77(2):139-48.

[198]  Hillyard DR, Monje VD, Mintz IM, *et al.* A new Conus peptide ligand for mammalian presynaptic $Ca^{2+}$ channels. Neuron 1992; 9(1):69-77.

[199]  Grantham CJ, Bowman D, Bath CP, Bell DC, Bleakman D. Omega-conotoxin MVIIC reversibly inhibits a human N-type calcium channel and calcium influx into chick synaptosomes. Neuropharmacology 1994; 33(2):255-8.

[200]  Vieira LB, Kushmerick C, Hildebrand ME, *et al.* Inhibition of high voltage-activated calcium channels by spider toxin PnTx3-6. J Pharmacol Exp Ther 2005; 314(3):1370-7.

[201]  Li-Smerin Y, Swartz KJ. Gating modifier toxins reveal a conserved structural motif in voltage-gated $Ca^{2+}$ and $K^+$ channels. Proc Natl Acad Sci U S A 1998; 95(15):8585-9.

[202]  Bourinet E, Stotz SC, Spaetgens RL, *et al.* Interaction of SNX482 with domains III and IV inhibits activation gating of alpha(1E) (Ca(V)2.3) calcium channels. Biophys J 2001; 81(1):79-88.

[203]  Chuang RS, Jaffe H, Cribbs L, Perez-Reyes E, Swartz KJ. Inhibition of T-type voltage-gated calcium channels by a new scorpion toxin. Nat Neurosci 1998; 1(8):668-74.

[204]  Edgerton GB, Blumenthal KM, Hanck DA. Inhibition of the activation pathway of the T-type calcium channel Ca(V)3.1 by ProTxII. Toxicon 2010; 56(4):624-36.

[205]  Ohkubo T, Yamazaki J, Kitamura K. Tarantula toxin ProTx-I differentiates between human T-type voltage-gated $Ca^{2+}$ Channels Cav3.1 and Cav3.2. J Pharmacol Sci 2010; 112(4):452-8.

[206]  Sidach SS, Mintz IM. Kurtoxin, a gating modifier of neuronal high- and low-threshold ca channels. J Neurosci 2002; 22(6):2023-34.

[207]  Mintz IM, Venema VJ, Swiderek KM, *et al.* P-type calcium channels blocked by the spider toxin omega-Aga-IVA. Nature 1992; 355(6363):827-9.

[208]  Winterfield JR, Swartz KJ. A hot spot for the interaction of gating modifier toxins with voltage-dependent ion channels. J Gen Physiol 2000; 116(5):637-44.

[209]  Newcomb R, Szoke B, Palma A, *et al.* Selective peptide antagonist of the class E calcium channel from the venom of the tarantula Hysterocrates gigas. Biochemistry 1998; 37(44):15353-62.

[210]  Arroyo G, Aldea M, Fuentealba J, Albillos A, Garcia AG. SNX482 selectively blocks P/Q $Ca^{2+}$ channels and delays the inactivation of $Na^+$ channels of chromaffin cells. Eur J Pharmacol 2003; 475(1-3):11-8.

[211]  Lampe RA, Defeo PA, Davison MD, *et al.* Isolation and pharmacological characterization of omega-grammotoxin SIA, a novel peptide inhibitor of neuronal voltage-sensitive calcium channel responses. Mol Pharmacol 1993; 44(2):451-60.

[212]  McDonough SI, Lampe RA, Keith RA, Bean BP. Voltage-dependent inhibition of N- and P-type calcium channels by the peptide toxin omega-grammotoxin-SIA. Mol Pharmacol 1997; 52(6):1095-104.

[213]  Kim JI, Takahashi M, Ohtake A, Wakamiya A, Sato K. Tyr13 is essential for the activity of omega-conotoxin MVIIA and GVIA, specific N-type calcium channel blockers. Biochem Biophys Res Commun 1995; 206(2):449-54.

[214] Flinn JP, Pallaghy PK, Lew MJ, *et al.* Roles of key functional groups in omega-conotoxin GVIA synthesis, structure and functional assay of selected peptide analogues. Eur J Biochem 1999; 262(2):447-55.

[215] Nadasdi L, Yamashiro D, Chung D, *et al.* Structure-activity analysis of a Conus peptide blocker of N-type neuronal calcium channels. Biochemistry 1995; 34(25):8076-81.

[216] Kim JI, Takahashi M, Ogura A, *et al.* Hydroxyl group of Tyr13 is essential for the activity of omega-conotoxin GVIA, a peptide toxin for N-type calcium channel. J Biol Chem 1994; 269(39):23876-8.

[217] Mould J, Yasuda T, Schroeder CI, *et al.* The alpha2delta auxiliary subunit reduces affinity of omega-conotoxins for recombinant N-type (Cav2.2) calcium channels. J Biol Chem 2004; 279(33):34705-14.

[218] Atkinson RA, Kieffer B, Dejaegere A, Sirockin F, Lefevre JF. Structural and dynamic characterization of omega-conotoxin MVIIA: the binding loop exhibits slow conformational exchange. Biochemistry 2000; 39(14):3908-19.

[219] Nielsen KJ, Schroeder T, Lewis R. Structure-activity relationships of omega-conotoxins at N-type voltage-sensitive calcium channels. J Mol Recognit 2000; 13(2):55-70.

[220] Schroeder CI, Smythe ML, Lewis RJ. Development of small molecules that mimic the binding of omega-conotoxins at the N-type voltage-gated calcium channel. Mol Divers 2004; 8(2):127-34.

[221] Schroeder CI, Doering CJ, Zamponi GW, Lewis RJ. N-type calcium channel blockers: novel therapeutics for the treatment of pain. Med Chem 2006; 2(5):535-43.

[222] Kim JI, Konishi S, Iwai H, *et al.* Three-dimensional solution structure of the calcium channel antagonist omega-agatoxin IVA: consensus molecular folding of calcium channel blockers. J Mol Biol 1995; 250(5):659-71.

[223] Takeuchi K, Park E, Lee C, *et al.* Solution structure of omega-grammotoxin SIA, a gating modifier of P/Q and N-type Ca$^{2+}$ channel. J Mol Biol 2002; 321(3):517-26.

[224] Lee CW, Bae C, Lee J, *et al.* Solution structure of kurtoxin: a gating modifier selective for Cav3 voltage-gated Ca$^{2+}$ channels. Biochemistry 2012; 51(9):1862-73.

[225] Snutch TP. Targeting chronic and neuropathic pain: the N-type calcium channel comes of age. NeuroRx 2005; 2(4):662-70.

[226] Yamada K, Teraoka T, Morita S, Hasegawa T, Nabeshima T. Omega-conotoxin GVIA protects against ischemia-induced neuronal death in the Mongolian gerbil but not against quinolinic acid-induced neurotoxicity in the rat. Neuropharmacology 1994; 33(2):251-4.

[227] Madden KP, Clark WM, Marcoux FW, *et al.* Treatment with conotoxin, an 'N-type' calcium channel blocker, in neuronal hypoxic-ischemic injury. Brain Res 1990; 537(1-2):256-62.

[228] Perez-Pinzon MA, Yenari MA, Sun GH, Kunis DM, Steinberg GK. SNX-111, a novel, presynaptic N-type calcium channel antagonist, is neuroprotective against focal cerebral ischemia in rabbits. J Neurol Sci 1997; 153(1):25-31.

[229] Omote K, Kawamata M, Satoh O, Iwasaki H, Namiki A. Spinal antinociceptive action of an N-Type voltage-dependent calcium channel blocker and the synergistic interaction with morphine. Anesthesiology 1996; 84(3):636-43.

[230] Suh HW, Song DK, Choi SR, Huh SO, Kim YH. Effects of intrathecal injection of nimodipine, omega-conotoxin GVIA, calmidazolium, and KN-62 on the antinociception induced by cold water swimming stress in the mouse. Brain Res 1997; 767(1):144-7.

[231]  Ino M, Yoshinaga T, Wakamori M, *et al.* Functional disorders of the sympathetic nervous system in mice lacking the alpha 1B subunit (Cav 2.2) of N-type calcium channels. Proc Natl Acad Sci U S A 2001; 98(9):5323-8.

[232]  Kim C, Jun K, Lee T, *et al.* Altered nociceptive response in mice deficient in the alpha(1B) subunit of the voltage-dependent calcium channel. Mol Cell Neurosci 2001; 18(2):235-45.

[233]  Saegusa H, Kurihara T, Zong S, *et al.* Suppression of inflammatory and neuropathic pain symptoms in mice lacking the N-type $Ca^{2+}$ channel. EMBO J 2001; 20(10):2349-56.

[234]  Saegusa H, Matsuda Y, Tanabe T. Effects of ablation of N- and R-type $Ca^{2+}$ channels on pain transmission. Neurosci Res 2002; 43(1):1-7.

[235]  Westenbroek RE, Hoskins L, Catterall WA. Localization of $Ca^{2+}$ channel subtypes on rat spinal motor neurons, interneurons, and nerve terminals. J Neurosci 1998; 18(16):6319-30.

[236]  Murakami M, Nakagawasai O, Suzuki T, *et al.* Antinociceptive effect of different types of calcium channel inhibitors and the distribution of various calcium channel alpha 1 subunits in the dorsal horn of spinal cord in mice. Brain Res 2004; 1024(1-2):122-9.

[237]  Pruneau D, Angus JA. Omega-conotoxin GVIA, the N-type calcium channel inhibitor, is sympatholytic but not vagolytic: consequences for hemodynamics and autonomic reflexes in conscious rabbits. J Cardiovasc Pharmacol 1990; 16(4):675-80.

[238]  Malmberg AB, Yaksh TL. Effect of continuous intrathecal infusion of omega-conopeptides, N-type calcium-channel blockers, on behavior and antinociception in the formalin and hot-plate tests in rats. Pain 1995; 60(1):83-90.

[239]  Staats PS, Yearwood T, Charapata SG, *et al.* Intrathecal ziconotide in the treatment of refractory pain in patients with cancer or AIDS: a randomized controlled trial. JAMA 2004; 291(1):63-70.

[240]  Scott DA, Wright CE, Angus JA. Actions of intrathecal omega-conotoxins CVID, GVIA, MVIIA, and morphine in acute and neuropathic pain in the rat. Eur J Pharmacol 2002; 451(3):279-86.

[241]  Cizkova D, Marsala J, Lukacova N, *et al.* Localization of N-type $Ca^{2+}$ channels in the rat spinal cord following chronic constrictive nerve injury. Exp Brain Res 2002; 147(4):456-63.

[242]  Newton RA, Bingham S, Case PC, Sanger GJ, Lawson SN. Dorsal root ganglion neurons show increased expression of the calcium channel alpha2delta-1 subunit following partial sciatic nerve injury. Brain Res Mol Brain Res 2001; 95(1-2):1-8.

[243]  Abe M, Kurihara T, Han W, Shinomiya K, Tanabe T. Changes in expression of voltage-dependent ion channel subunits in dorsal root ganglia of rats with radicular injury and pain. Spine (Phila Pa 1976) 2002; 27(14):1517-24; discussion 25.

[244]  Wang YX, Gao D, Pettus M, Phillips C, Bowersox SS. Interactions of intrathecally administered ziconotide, a selective blocker of neuronal N-type voltage-sensitive calcium channels, with morphine on nociception in rats. Pain 2000; 84(2-3):271-81.

[245]  Penn RD, Paice JA. Adverse effects associated with the intrathecal administration of ziconotide. Pain 2000; 85(1-2):291-6.

[246]  Takahara A, Koganei H, Takeda T, Iwata S. Antisympathetic and hemodynamic property of a dual L/N-type $Ca^{2+}$ channel blocker cilnidipine in rats. Eur J Pharmacol 2002; 434(1-2):43-7.

[247]  Williams JA, Day M, Heavner JE. Ziconotide: an update and review. Expert Opin Pharmacother 2008; 9(9):1575-83.

[248] Pajouhesh H, Feng ZP, Zhang L, *et al.* Structure-activity relationships of trimethoxybenzyl piperazine N-type calcium channel inhibitors. Bioorg Med Chem Lett 2012; 22(12):4153-8.

[249] Jurkat-Rott K, Lehmann-Horn F. The impact of splice isoforms on voltage-gated calcium channel alpha1 subunits. J Physiol 2004; 554(Pt 3):609-19.

[250] Altier C, Dale CS, Kisilevsky AE, *et al.* Differential role of N-type calcium channel splice isoforms in pain. J Neurosci 2007; 27(24):6363-73.

[251] Zamponi GW, Bourinet E, Nelson D, Nargeot J, Snutch TP. Crosstalk between G proteins and protein kinase C mediated by the calcium channel alpha1 subunit. Nature 1997; 385(6615):442-6.

[252] McDavid S, Currie KP. G-proteins modulate cumulative inactivation of N-type (Cav2.2) calcium channels. J Neurosci 2006; 26(51):13373-83.

[253] Kolosov A, Goodchild CS, Cooke I. CNSB004 (Leconotide) causes antihyperalgesia without side effects when given intravenously: a comparison with ziconotide in a rat model of diabetic neuropathic pain. Pain Med 2010; 11(2):262-73.

[254] Kolosov A, Aurini L, Williams ED, Cooke I, Goodchild CS. Intravenous injection of leconotide, an omega conotoxin: synergistic antihyperalgesic effects with morphine in a rat model of bone cancer pain. Pain Med 2011; 12(6):923-41.

[255] Pexton T, Moeller-Bertram T, Schilling JM, Wallace MS. Targeting voltage-gated calcium channels for the treatment of neuropathic pain: a review of drug development. Expert Opin Investig Drugs 2011; 20(9):1277-84.

[256] Wright CE, Robertson AD, Whorlow SL, Angus JA. Cardiovascular and autonomic effects of omega-conotoxins MVIIA and CVID in conscious rabbits and isolated tissue assays. Br J Pharmacol 2000; 131(7):1325-36.

[257] Pietrobon D. CaV2.1 channelopathies. Pflugers Arch 2010; 460(2):375-93.

[258] Eikermann-Haerter K, Dilekoz E, Kudo C, *et al.* Genetic and hormonal factors modulate spreading depression and transient hemiparesis in mouse models of familial hemiplegic migraine type 1. J Clin Invest 2009; 119(1):99-109.

[259] van den Maagdenberg AM, Pizzorusso T, Kaja S, *et al.* High cortical spreading depression susceptibility and migraine-associated symptoms in Ca(v)2.1 S218L mice. Ann Neurol 2010; 67(1):85-98.

[260] de Souza AH, Lima MC, Drewes CC, *et al.* Antiallodynic effect and side effects of Phalpha1beta, a neurotoxin from the spider Phoneutria nigriventer: comparison with omega-conotoxin MVIIA and morphine. Toxicon 2011; 58(8):626-33.

[261] Chen CC, Lamping KG, Nuno DW, *et al.* Abnormal coronary function in mice deficient in alpha1H T-type $Ca^{2+}$ channels. Science 2003; 302(5649):1416-8.

[262] Kim D, Park D, Choi S, *et al.* Thalamic control of visceral nociception mediated by T-type $Ca^{2+}$ channels. Science 2003; 302(5642):117-9.

[263] Ikeda H, Heinke B, Ruscheweyh R, Sandkuhler J. Synaptic plasticity in spinal lamina I projection neurons that mediate hyperalgesia. Science 2003; 299(5610):1237-40.

[264] Choi S, Na HS, Kim J, *et al.* Attenuated pain responses in mice lacking Ca(V)3.2 T-type channels. Genes Brain Behav 2007; 6(5):425-31.

[265] Nelson MT, Woo J, Kang HW, *et al.* Reducing agents sensitize C-type nociceptors by relieving high-affinity zinc inhibition of T-type calcium channels. J Neurosci 2007; 27(31):8250-60.

[266] Kim D, Song I, Keum S, *et al.* Lack of the burst firing of thalamocortical relay neurons and resistance to absence seizures in mice lacking alpha(1G) T-type $Ca^{2+}$ channels. Neuron 2001; 31(1):35-45.

[267]   Song I, Kim D, Choi S, *et al.* Role of the alpha1G T-type calcium channel in spontaneous absence seizures in mutant mice. J Neurosci 2004; 24(22):5249-57.

[268]   Khosravani H, Zamponi GW. Voltage-gated calcium channels and idiopathic generalized epilepsies. Physiol Rev 2006; 86(3):941-66.

[269]   Heron SE, Khosravani H, Varela D, *et al.* Extended spectrum of idiopathic generalized epilepsies associated with CACNA1H functional variants. Ann Neurol 2007; 62(6):560-8.

[270]   Khosravani H, Altier C, Simms B, *et al.* Gating effects of mutations in the Cav3.2 T-type calcium channel associated with childhood absence epilepsy. J Biol Chem 2004; 279(11):9681-4.

[271]   Peloquin JB, Khosravani H, Barr W, *et al.* Functional analysis of Ca3.2 T-type calcium channel mutations linked to childhood absence epilepsy. Epilepsia 2006; 47(3):655-8.

[272]   Vitko I, Chen Y, Arias JM, *et al.* Functional characterization and neuronal modeling of the effects of childhood absence epilepsy variants of CACNA1H, a T-type calcium channel. J Neurosci 2005; 25(19):4844-55.

[273]   Splawski I, Yoo DS, Stotz SC, *et al.* CACNA1H mutations in autism spectrum disorders. J Biol Chem 2006; 281(31):22085-91.

[274]   Anderson MP, Mochizuki T, Xie J, *et al.* Thalamic Cav3.1 T-type $Ca^{2+}$ channel plays a crucial role in stabilizing sleep. Proc Natl Acad Sci U S A 2005; 102(5):1743-8.

[275]   Lee J, Kim D, Shin HS. Lack of delta waves and sleep disturbances during non-rapid eye movement sleep in mice lacking alpha1G-subunit of T-type calcium channels. Proc Natl Acad Sci U S A 2004; 101(52):18195-9.

[276]   Kraus RL, Li Y, Gregan Y, *et al. In vitro* characterization of T-type calcium channel antagonist TTA-A2 and *in vivo* effects on arousal in mice. J Pharmacol Exp Ther 2010; 335(2):409-17.

[277]   Takahashi H, Yoshimoto M, Higuchi H, Shimizu T, Hishikawa Y. Different effects of L-type and T-type calcium channel blockers on the hypnotic potency of triazolam and zolpidem in rats. Eur Neuropsychopharmacol 1999; 9(4):317-21.

[278]   Uebele VN, Gotter AL, Nuss CE, *et al.* Antagonism of T-type calcium channels inhibits high-fat diet-induced weight gain in mice. J Clin Invest 2009; 119(6):1659-67.

[279]   Todorovic SM, Jevtovic-Todorovic V. T-type voltage-gated calcium channels as targets for the development of novel pain therapies. Br J Pharmacol 2011; 163(3):484-95.

[280]   Feher M, Schmidt JM. Property distributions: differences between drugs, natural products, and molecules from combinatorial chemistry. J Chem Inf Comput Sci 2003; 43(1):218-27.

[281]   Olivera BM. Conus peptides: biodiversity-based discovery and exogenomics. J Biol Chem 2006; 281(42):31173-7.

[282]   Buczek O, Bulaj G, Olivera BM. Conotoxins and the posttranslational modification of secreted gene products. Cell Mol Life Sci 2005; 62(24):3067-79.

[283]   Possani LD, Becerril B, Delepierre M, Tytgat J. Scorpion toxins specific for $Na^+$ channels. Eur J Biochem 1999; 264(2):287-300.

[284]   Pi C, Liu Y, Peng C, *et al.* Analysis of expressed sequence tags from the venom ducts of Conus striatus: focusing on the expression profile of conotoxins. Biochimie 2006; 88(2):131-40.

[285]   Bhatia S, Kil YJ, Ueberheide B, *et al.* Constrained *De Novo* Sequencing of Conotoxins. J Proteome Res 2012.

[286]   Pi C, Liu J, Peng C, *et al.* Diversity and evolution of conotoxins based on gene expression profiling of Conus litteratus. Genomics 2006; 88(6):809-19.

[287] Violette A, Biass D, Dutertre S, *et al.* Large-scale discovery of conopeptides and conoproteins in the injectable venom of a fish-hunting cone snail using a combined proteomic and transcriptomic approach. J Proteomics 2012.

[288] Kaas Q, Yu R, Jin AH, Dutertre S, Craik DJ. ConoServer: updated content, knowledge, and discovery tools in the conopeptide database. Nucleic Acids Res 2012; 40(Database issue):D325-30.

[289] Herzig V, Wood DL, Newell F, *et al.* ArachnoServer 2.0, an updated online resource for spider toxin sequences and structures. Nucleic Acids Res 2011; 39(Database issue):D653-7.

[290] Srinivasan KN, Gopalakrishnakone P, Tan PT, *et al.* SCORPION, a molecular database of scorpion toxins. Toxicon 2002; 40(1):23-31.

[291] Tan PT, Veeramani A, Srinivasan KN, Ranganathan S, Brusic V. SCORPION2: a database for structure-function analysis of scorpion toxins. Toxicon 2006; 47(3):356-63.

[292] Pedersen SL, Tofteng AP, Malik L, Jensen KJ. Microwave heating in solid-phase peptide synthesis. Chem Soc Rev 2012; 41(5):1826-44.

[293] Coin I, Beyermann M, Bienert M. Solid-phase peptide synthesis: from standard procedures to the synthesis of difficult sequences. Nat Protoc 2007; 2(12):3247-56.

[294] Gowd KH, Yarotskyy V, Elmslie KS, *et al.* Site-specific effects of diselenide bridges on the oxidative folding of a cystine knot peptide, omega-selenoconotoxin GVIA. Biochemistry 2010; 49(12):2741-52.

[295] Qoronfleh MW, Hesterberg LK, Seefeldt MB. Confronting high-throughput protein refolding using high pressure and solution screens. Protein Expr Purif 2007; 55(2):209-24.

[296] Swensen AM, Niforatos W, Vortherms TA, *et al.* An Automated Electrophysiological Assay for Differentiating Ca(V)2.2 Inhibitors Based on State Dependence and Kinetics. Assay Drug Dev Technol 2012.

[297] Swensen AM, Herrington J, Bugianesi RM, *et al.* Characterization of the substituted N-triazole oxindole TROX-1, a small-molecule, state-dependent inhibitor of Ca(V)2 calcium channels. Mol Pharmacol 2012; 81(3):488-97.

[298] Miwa JM, Ibanez-Tallon I, Crabtree GW, *et al.* lynx1, an endogenous toxin-like modulator of nicotinic acetylcholine receptors in the mammalian CNS. Neuron 1999; 23(1):105-14.

[299] Ibanez-Tallon I, Wen H, Miwa JM, *et al.* Tethering naturally occurring peptide toxins for cell-autonomous modulation of ion channels and receptors *in vivo*. Neuron 2004; 43(3):305-11.

[300] Wu Y, Cao G, Pavlicek B, Luo X, Nitabach MN. Phase coupling of a circadian neuropeptide with rest/activity rhythms detected using a membrane-tethered spider toxin. PLoS Biol 2008; 6(11):e273.

[301] Derman AI, Prinz WA, Belin D, Beckwith J. Mutations that allow disulfide bond formation in the cytoplasm of Escherichia coli. Science 1993; 262(5140):1744-7.

[302] Ibanez-Tallon I, Nitabach MN. Tethering toxins and peptide ligands for modulation of neuronal function. Curr Opin Neurobiol 2012; 22(1):72-8.

[303] McCleskey EW. Neuroscience: a local route to pain relief. Nature 2007; 449(7162):545-6.

Send Orders for Reprints to reprints@benthamscience.net

# CHAPTER 5

# Drugs Targeting Microglial Activation in Hypoxic Damage to the Developing White Matter

Gurugirijha Rathnasamy, Madhuvika Murugan, Eng-Ang Ling and Charanjit Kaur*

*Department of Anatomy, MD10, 4 Medical Drive, Yong Loo Lin School of Medicine, National University of Singapore, Singapore-117597, Singapore*

**Abstract:** Advancements in medical facilities have increased the survival rates of preterm infants, yet the propensity of these infants developing white matter (WM) damage (WMD) remains largely unaltered. It is estimated that 4-10% of premature newborns are affected by WMD which leads to severe long term neurological consequences such as cerebral palsy, epilepsy, hearing and vision impairments. Currently, there is no effective treatment for WMD and all available therapies target secondary pathologies that develop as a result of damage to white matter tissue. WMD may occur as a result of a number of causes, with hypoxia being one of the major underlying factors. The multi-factorial aetiology of WMD further hinders the development of specific drugs for the treatment of this clinical condition. The pathological hallmarks include death of oligodendrocytes, degeneration of axons, hypomyelination, microglial activation and astrogliosis. Although role of microglia is considered to be neuroprotective in many neurological conditions, our recent studies have shown that microglial activation is an important event in WMD, the consequences of which include increased release of pro-inflammatory cytokines, proteinases and glutamate, and generation of reactive oxygen intermediates, along with undesirable sequestration of excess iron released during hypoxia. Hence amelioration of microglial activation has been considered as the key target in the therapy of WMD. Drugs such as-glutamate receptor antagonists, iron chelators, antioxidants, ion channel blockers and certain immunosuppressive agents that target microglial activation have been identified, but their mechanism of action still remains elusive. This chapter will summarize the currently available and potential drugs along with their mechanism of action underlying the suppression of microglial activation for the treatment of WMD in the developing brain.

**Keywords:** Developing brain, hypoxia, white matter damage, microglia, inflammation, excitotoxicity, oxidative stress, glutamate receptor antagonist, iron chelators, anti-inflammatory drugs, antioxidants.

*Address correspondence to Charanjit Kaur: Department of Anatomy, MD10, 4 Medical Drive, Yong Loo Lin School of Medicine, National University of Singapore, Singapore-117597, Singapore; Tel: 65-65163209; Fax: 65-67787643; E-mail: antkaurc@nus.edu.sg

Atta-ur-Rahman (Ed)
All rights reserved-© 2013 Bentham Science Publishers

# INTRODUCTION

Preterm birth is the leading cause of death in infants within the first four weeks of life [1]. According to the 2012 World Health Organization report, an estimated 15 million premature infants are born annually, of which over one million children die due to complications of preterm birth. The surviving premature infants are at increased risk of suffering brain damage, particularly white matter (WM) damage (WMD) in infants born at a gestational age of 24-32 weeks [2-7]. It is estimated that 4-10% of premature newborns suffer from WMD with long-term neurological impairments such as cerebral palsy, epilepsy, learning disabilities, hearing and vision impairments [5, 8].

Evidence points to reduced supply of oxygen to the brain or hypoxia, to be one of the most important factors causing WMD in premature neonates [9-11]. The pathological hallmarks of hypoxic WMD include death of oligodendrocytes, damage to axons, delayed/disrupted myelination, microglial activation and astrogliosis. A comparative study showed that, of the glial cells, oligodendrocytes are selectively vulnerable to hypoxic insults [12]. During development, the oligodendrocytes lack certain antioxidant enzymes which render them highly susceptible to oxidative stress [13]. The death of oligodendrocytes could also be triggered by excess excitatory amino acids, trophic factor deprivation, and activation of apoptotic pathways in hypoxic/ishemic injuries [14]. Hypoxia induces activation of microglia initiating a myriad of molecular events such as nitric oxide (NO) production, oxygen radicals formation, iron sequestration, disruption of calcium regulation and release of pro-inflammatory cytokines, proteases, neurotransmitters, and excitatory amino acids, primarily glutamate [15, 16]. In light of the above and given the abundance of activated/amoeboid microglia in the developing white matter, microglial activation has been thought to be a key mediator in causing damage to oligodendrocytes and axons in hypoxic WM.

To date, there are no specific therapies for the treatment of neonatal WMD and all available drugs are for symptomatic relief, that is, they are prescribed in response to secondary pathologies. The multifaceted aetiology of hypoxic WMD has complicated the process of drug discovery. However, amelioration of microglial activation is thought to be a potent therapeutic target in reducing the hypoxia-induced damage to the WM [17]. Several drugs including glutamate receptor antagonists, iron chelators, antioxidants, ion channel blockers and certain immunosuppressive agents have been identified to suppress microglial activation [18-21]. These pharmacological agents act through various modes in order to attenuate activation of microglia. However, for some of these drugs, the mechanism of action still remains elusive. Certain microglial suppressors used in the clinics for the treatment of WMD [18, 22] will be discussed in detail in this chapter.

## PATHOLOGY OF WHITE MATTER DAMAGE

In response to hypoxic-ischemic insults the pathological consequences in the developing WM composed of glial cells (microglia, astrocytes and oligodendrocytes) and unmyelinated axons [23], range from necrosis to myelination deficits. Hallmark events include oligodendroglial damage, axonal swelling, microglial activation and astrocytosis [24-26]. Following a hypoxic-ischemic insult, necrotic and apoptotic cells were visualized in the WM [27, 28] and these cells were identified to be oligodendrocytes [28]. Besides apoptotic and necrotic oligodendrocytes, glioblasts known to be the oligodendrocyte precursors [29] exhibited the splitting of inner and outer nuclear membrane with ballooning of the nuclear envelope. Along with the apoptotic oligodendrocytes, the axons, which are essential for transmission of neural information across and within the cerebral hemispheres, were found to be swollen and degenerating [30].

The axons in the developing human brain are unmyelinated and the process of myelination in brain starts during the mid-gestation period and continues through the first two years of life [31]. The myelin sheaths which favor neural conduction/saltatory conduction of action potential along the axons are extensions of oligodendrocyte processes. Oligodendrocyte precursors are known to dominate the WM during 23-32 weeks of gestation in humans [32]. Hence, a hypoxic-ischemic injury during this gestational period, which causes apoptosis of oligodendrocytes results in delayed/disrupted myelin formation leading to severe neurobehavioral anomalies that last through childhood and adulthood [5]. Reduced myelin basic protein expression has been reported in the WM following a hypoxic-ischemic injury [30, 33, 34]. The cystic cavities found during the later stages of WM injury are usually accompanied by impaired or delayed myelination [35-37]. We have reported that in response to hypoxia, there was a significant increase in the number of degenerating axons in the neonatal brain [30]. The affected axons in hypoxic WM lacked axoplasm and the few axons that were myelinated had distorted myelin sheath [30].

The presence of activated microglia has been detected in the WM of children with signs of periventricular leukomalacia [38-40] and in asphyxiated animal foetuses [41]. Microglia activated by a hypoxic insult, are known to generate reactive oxygen/nitrogen species [20], which are involved in the pathogenetic cascade of WMD [42]. Along with this, the inflammatory reactions resulting in oligodendrocyte progenitor death are attributed to the abundance of microglia in the developing WM [43]. Furthermore, in WMD, microglial cells were observed to phagocytose the degenerating axons, necrotic and apoptotic cells [17].

Astrocytosis is another characteristic feature of WMD [44]. Though astrocytosis could favor in the repair of WM, prolonged astrocytosis leads to damage of the axons and oligodendrocytes [45], either through production of pro-inflammatory cytokines [46] or by inhibiting axonal re-growth [47]. The astrocytes along with the extracellular matrix form the glial scar at the site of lesion following brain injury [48]. In addition, astrocyte processes are known to be closely associated with blood vessels and are known to produce vascular endothelial growth factor (VEGF), which induces vasculogenesis. In developing brain, in response to chronic sublethal hypoxia, VEGF was found to be up-regulated in astrocytes and was also implicated in enhancing the vascular permeability [49]. The up-regulation of VEGF expression in astrocytes in WM, in response to hypoxia was suggested to have a critical role in edema formation by enhancing vascular leakage [30].

The other mediators of WMD include factors such as circulatory disturbances, edema and hemorrhage with consequent coagulation necrosis [50]. Germinal matrix hemorrhage [51] and intraventricular hemorrhage often accompany WMD [52]. Following hypoxic injury the ependymal cells lining the lateral ventricles possessed vacuoles and the intercellular spaces were widened [17]. These structural changes could lead to extravasation of cerebro-spinal fluid into the periventricular WM thereby enhancing edema formation [17]. Studies have also reported the widening of perivascular spaces and consequent edema formation in WM [30, 53, 54]. Some authors have reported that the vulnerability of WM to hypoxic injury may be due to the hypo-vascularisation of this region [7].

## MICROGLIA IN THE WM

Microglia, the primary immune effector cells in the central nervous system (CNS) are ubiquitously distributed throughout the brain and spinal cord. Evidence suggests that microglia in the white and grey matter in the CNS are phenotypically distinct [55]. In support of this theory, it has been observed that microglia in the WM of mice have a constitutively higher level of expression of co-stimulatory molecules (molecules that have the ability to modulate microglial response) such as B7.2 (CD86) than their cortical counterparts. This study suggests that microglia residing in WM exist with a higher basal level of activation under physiological conditions [56]. Also it has been shown in humans, monkeys, rats and dogs, that the number of major histocompatibility complex II (MHC II)-positive microglia under physiological conditions, is higher in WM (corpus callosum, internal capsule) than in the cerebral cortex [57-63]. In addition, Tim-3, a membrane protein involved in the regulation of macrophage and T-cell responses, was found to be expressed only on corpus callosal microglia in the non-inflamed human brain [64]. WM microglia were shown to be more sensitive to age-related increase in the basal level of activation than their grey matter counterparts [59, 62, 65]. Furthermore, microglia activation was mainly observed in the WM of rats in an experimental model of chronic cerebral hypoperfusion [66-68]. In spite of the accumulating evidence on the phenotypic differences between white and grey matter microglia, its functional significance remains unclear.

Microglia also exhibit phenotypic variation during development. In the developing brain, microglia possess an amoeboid morphology and are referred to as amoeboid microglial cells (AMC). In the adult brain the microglia are in a quiescent state, with ramified morphology and hence are termed ramified microglial cells (RMC). Under pathological conditions such as trauma, infection and hypoxia, RMC retract their processes and assume an amoeboid form [69, 70]. The AMC are found mainly within the WM areas in the perinatal brain [71]. Several studies performed on rodents suggest that AMC are the precursors of RMC [72], especially because the decline in AMC numbers with age occurs simultaneously with an increase in number of RMC throughout the brain, including WM regions [73, 74]. A recent study compared the gene expression profile of AMC and RMC present in the corpus callosum of rat brain, and identified that the AMC exhibited increased expression of genes essential for migration, proliferation, apoptosis and phagocytosis [75]. In light of the above, the AMC are considered to be in an activated state and this is supported by a study that showed binding of CD68, an activated microglia marker, to AMC during late gestational period in humans [43]. This study also reported a transient overabundance of AMC in WM during prenatal period, suggesting that these cells may play a vital role in brain development processes, such as, vascularization [76], involution of the germinal matrix [77], programmed cell death [78], axonal development [79], and myelination [80]. The AMC are also involved in "pruning" of unwanted axons and synapses in the developing brain [81-83], with their high content of hydrolytic enzymes [84] enabling them to phagocytose debris including dead cells, degenerating axons and redundant synapses [85, 86].

Although microglia have a beneficial role under physiological conditions, their prolonged activation may lead to the release of toxic substances that cause bystander damage to the surrounding cells in the neural parenchyma. The microglia are sensitive to minute changes in the environment and are activated by stimuli such as infection, hypoxia-ischemia, lipopolysaccharide (LPS), amyloid β, and changes in extracellular levels of potassium [87-90]. However, in the immature WM, microglial activation due to a hypoxic insult is the most potent mediator of WMD, the major pathological substrate of cerebral palsy in the premature infant. Increased density of AMC in the cerebral WM during the peak window of vulnerability for WMD [43], has made microglial activation an important suspect in the pathogenesis of WMD.

## CONSEQUENCES OF MICROGLIAL ACTIVATION IN HYPOXIC WM IN THE DEVELOPING BRAIN

The state of microglia in the WM is governed by their external environment and is typically characterized as "quiescent" or "activated". Apart from these two states, these cells exhibit an array of intermediate morphologies and are termed as "alternatively activated" [91-93]. Depending upon their state of activation, microglia display a repertoire of activity including morphologic/phenotypic changes, increased gene expression, production of cytokines, release of glutamate, sequestration of iron, generation of ROS, production of NO, secretion of proteases and phagocytosis (See review by [94]). Since microglial activation has been implicated in the pathogenesis of hypoxic WMD, understanding the consequences of microglial activation, therefore, is critical in the development of novel therapies to prevent the progression of damage. In this section, we summarize emerging concepts from animal and human models concerning the regulation of microglial activity in response to hypoxia (Fig. **1**) and the pathways by which they mediate death of oligodendrocytes and delayed myelination thus contributing to WMD in neonates.

## Morphological Changes

Following a hypoxic insult in the adult brain, microglia become activated, that is, from a ramified resting state they acquire a more amoeboid phenotype with an enlarged cell body and stout processes. However, in the developing brain, the microglia are already in the amoeboid form, hence a significant change in its phenotypic features following injury was not observed [95]. In addition, transitional forms between normal ameboid and activated-like microglial cells were identified at the site of injury, supporting the view that ameboid microglia become activated [95]. However, the response of immature (ameboid) microglia to injury in the developing CNS has received little attention and needs to be investigated.

## Phagocytosis

Phagocytic microglia are considered to be the maximal immune responsive state of microglia. The phagocytic property of microglia, such as uptake of exogenous substances, activated surface receptors, localisation of antigens and hydrolytic enzymes associated with phagocytosis, has been reported under various experimental conditions [96]. Ultrastructural studies have reported the phagocytic nature of AMC in neonatal rat brain following injection of *Escherichia coli* (*E. coli*), a bacteria [97]. In this study, it was observed that AMC were actively engaged in phagocytosing *E. coli* and clearance of most *E. coli* was achieved within 3 h of injection [97]. Also following a hypoxic insult, phagocytic AMC were identified in the WM of neonatal rat brain [27]. These cells were engaged in the phagocytosis of degenerating axons and apoptotic cells [27, 30, 98]; and the apoptotic cells in the hypoxic WM were identified to be oligodendrocytes [46]. Consistent with these studies, another study showed that AMC in the fetal brain engulfed apoptotic and necrotic cells following transient maternal hypoxia [99]. Based on these studies it is postulated that AMC in the developing WM serve as a protective barrier under physiological conditions, and since the blood brain barrier (BBB) is not fully functional they serve as scavengers during pathological conditions.

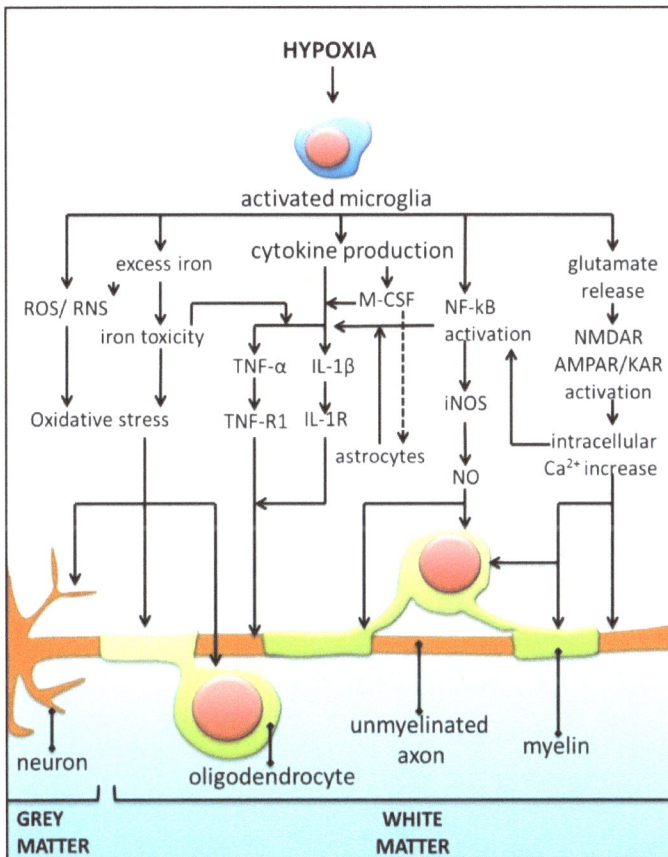

**Figure 1:** Schematic diagram showing the consequences of microglial activation in the white matter. Microglia upon activation releases glutamate and cytokines. The released glutamate in turn activates the microglial glutamate receptors (NMDAR/AMPAR/KAR) resulting in increased intracellular Ca2+ levels. Following this NF-κB is activated and leads to iNOS and NO production resulting in WMD. The major cytokines released by microglia namely TNF-α and IL-1β cause WMD by binding to their receptors TNF-R1 and IL-1R, respectively, expressed on the axons and oligodendrocytes of WM. Additionally M-CSF secreted by microglia binds to their receptors expressed on astrocytes resulting in enhanced production of cytokines. Activated microglia also accumulate excess iron which leads to oxidative stress through enhanced generation of free radicals.

## Antigen Presentation

AMCs constitutively express MHC I antigens [100], which are required for antigen presentation. Antigen presentation is an immunological process by which macrophages capture antigens and then enable their recognition by T-cells. Under physiological conditions, the expression of MHC antigens in the brain is restricted to microglia [101] and was reported to enhance their phagocytic activity [100]. Also the expression of MHC I antigens on AMC facilitated interaction with infiltrating lymphocytes, owing to the immature BBB during early development [96]. MHC II, however, are not constitutively expressed by AMC. The expression of MHC II has been reported only in the activated microglia, *e.g.,* –in response to LPS [102] or interferon-γ (IFN-γ) [103] or *E. coli* administration [97] or under hypoxic-ischemic conditions [104, 105]. It has been reported that hypoxia induces microglial activation with dramatic increase in immunoreactivity for MHCI/II antigens in the perinatal brain [105]. This is consistent with studies conducted in adult models of middle cerebral artery occlusion [106] and global ischemia [107-109]. In the hypoxic neonatal brain, the increase in MHC II expression in microglia was found to be transient [105] and this may be due to the immunological immaturity and/or developmental plasticity of the immunocompetent cells in the brain [110]. Increasing evidence suggests that up-regulation of MHC antigens in microglia is accompanied by co-stimulatory molecules such as CD11a, CD54, CD58, CD80 and CD 86 under hypoxic-ischemic conditions. These co-stimulatory molecules are essential for optimal antigen presentation and stimulation of antigen specific T lymphocytes [111]. The microglial expression of MHC I/II antigens along with the co-stimulatory molecules following hypoxic insults suggests that they have the capability of interacting with helper/inducer cells to mount a potential immune response during hypoxic WMD.

## Cytokine Production

Hypoxia-induced microglial activation is known to be associated with upregulated expression of cytokines. Cytokines are immunomodulatory signaling molecules that are extensively involved in intercellular communication. In the brain, microglia serve as the predominant source of cytokines [96, 112], with astrocytes being the secondary source. Cytokines produced by glial cells are one of the major contributors of damage to WM following perinatal hypoxia (See review by [113]).

In neonatal rats, hypoxia induced activation of microglial cells resulted in the production of inflammatory cytokines such as tumor necrosis factor (TNF)-α and interleukin (IL)-1β *via* mitogen activated protein kinase (MAPK)/nuclear factor-kB (NF-kB) signaling pathway [19, 28]. In the developing WM, hypoxia-induced sequestration of iron and activation of microglial N-methyl D-aspartate receptors (NMDAR) enhanced the release of pro-inflammatory mediators such as TNF-α and IL-1β [19, 20]. Astrocytes are also known to produce cytokines in hypoxic insults [114, 115]. However, it has been shown that microglial cytokines orchestrate astrocytic responses. For instance, microglial colony stimulating factor-1 (CSF-1) *via* astrocytic CSF-1 receptor activates the synthesis of TNF-α and IL-1β by the latter cell type [15].

Cytokines bind to their corresponding receptor triggering a cascade of signaling pathways in the target cell. Increased expression of cytokine receptors on oligodendrocytes has been reported following brain injury, which make these cells the potential target for damage [116-118]. Under hypoxic conditions, activated microglia are known to produce inflammatory cytokines such as TNF-α and IL-1β [28]. TNF-α enables it's signaling by binding to its receptors tumor necrosis factor receptor (TNF-R) 1or 2, and IL-1β acts through its receptors interleukin-1 receptor (IL-1R) 1 or 2. TNF-R1 and IL-1R1 are demonstrated to be expressed on oligodendrocytes and axons resulting in WM lesions [46]. Binding of TNF-α to the TNF-R1 activates programmed cell death signaling pathway resulting in apoptosis of oligodendrocytes [119]. TNF/TNF-R1 signaling is also associated with the activation of myelin associated sphingomyelinase (SMase). Activation of SMase results in breakdown of membrane sphingomyelin causing accumulation of ceramide, a signal that induces mitochondrial-mediated apoptosis [120]. However, IL-1β, unlike TNF-α, is not toxic to oligodendrocytes. IL-1β blocks oligodendrocyte proliferation at the late oligodendrocyte progenitor stage [121] leading to reduced number of cells. Put together, these findings suggest that TNF/TNF-1R and IL-1β/IL-1R1 signaling might be involved in oligodendrocyte apoptosis or reduced proliferation and subsequently delayed myelination [46, 122] contributing to hypoxic WMD.

Apart from exerting their effect on oligodendrocytes, it has been hypothesized that cytokines such as TNF-α and lymphotoxin (LT)-α may cause direct damage to myelin

[123] by mediating lipid changes within the myelin. Cytokine-induced activation of SMase resulted in the accumulation of ceramide and cholesteryl esters [124]. Cholesteryl-ester accumulation in myelin may expose antigenic sites in myelin rendering it vulnerable to attack by invading and resident immune cells [125].

Enhanced expression of chemokines and their receptors by activated microglia has been identified. Chemokines are chemotactic cytokines that mediate infiltration of leukocytes to the vicinity of injury. Upon stimulation by LPS, increased levels of chemokines such as interferon-γ-inducible protein-10, regulated on activation normal T expressed and secreted (RANTES) and macrophage inflammatory protein-1α (MIP-1α), as well as their receptors CXCR3 and CXCR5 were detected in activated microglia [126]. Hypoxia stimulated secretion of stromal cell-derived factor-1α (SDF-1α) and promoted migration of microglia by inducing the expression of CXCR4 on them [127]. In certain WM diseases, excess production of chemokines such as MIP-1α/RANTES and fractalkine and their corresponding receptors have been observed in actively demyelinating plaques. These chemokines were selectively expressed on microglia and monocyte-derived macrophages [128], indicating the infiltration of monocytes to the site of damage. In addition, it has been reported that chemokines enhance migration of microglia [129]. Infiltrating cells of the immune system and activated glial cells (astrocytes and microglia) express a wide range of chemokines [130-136]. Microglia produce chemokines such as monocyte chemoattractant protein-1 (MCP-1), MIP-1α, MIP-1β, MIP-2 and RANTES when stimulated by hypoxia [128]. These chemokines regulate the migration of microglia, monocytes and lymphocytes to the site of damage [137, 138] aggravating the inflammatory response in WMD following hypoxia.

## Release of Glutamate

Glutamate is the most predominant excitatory neurotransmitter in the brain. Several animal models of hypoxia/ischemia have implicated glutamate to be a critical mediator in the pathogenesis of WMD. Elevated glutamate levels were measured by *in vivo* microdialysis in neonatal rat brain following hypoxic insults [139, 140]. Another study reported an enhancement in glutamate levels in the CSF of term infants following perinatal hypoxia [141]. Glutamate production by microglia was increased in response to treatment with LPS and viral infection [142]. Activated microglia are known to contribute to the excess glutamate levels in the WM following hypoxic insults [143-145]. Perinatal hypoxia also alters the uptake of glutamate by astrocytic glutamate transporters, the primary mediators of glutamate homeostasis, contributing to the increased levels of extracellular glutamate [146]. Excess glutamate has been shown to be toxic to oligodendroglia *in vivo* and *in vitro* by receptor-independent [147-149] and receptor-mediated mechanisms [148, 150-153] resulting in WMD.

## Activation of Glutamate Receptors

Release of glutamate has been reported in the WM following hypoxia as discussed in the previous subsection. Excess glutamate acts through activation of its receptors-both ionotropic (α-amino-3-hydroxy-5-methyl-4-isoxazole-propionic acid receptors [AMPAR], NMDAR, kainate recceptors [KAR]) and metabotropic (mGluR), resulting in receptor-mediated excitotoxicity, a cardinal mediator of oligodendrocyte death in developing WM in hypoxic ischemic injuries [6, 154, 155].

Functional ionotropic glutamate receptors identified in cultured rat microglia mainly include AMPAR (GluRs 1–3, mainly in the flip form) and KAR [156, 157]. Addition of glutamate or kainate to microglial cultures triggered TNFα release which was prevented by cyclothiazide, a blocker of AMPAR [157]. Increased glutamate in response to hypoxia is known to instigate the production of cytokines by activated microglia [143]. Previous studies have reported the expression of AMPAR and KAR in microglia and that their activation enhances the production of TNF-α along with other cytokines causing death of oligodendrocytes [156, 158]. In addition, it was demonstrated that administration of exogenous glutamate led to the release of microglial TNF-α and IL-1β *via* activation of AMPARs [143]. Recently, functional NMDAR were reported in microglia located in the WM following a hypoxic insult [19]. Another study on transient forebrain ischemia in rats showed the expression of NR1 subunit of the NMDAR in microglia [151, 159]. Hypoxia-induced activation of microglial NMDAR mediated death of oligodendrocytes *via* the production of NO [19]. Also in this study it was shown that hypoxia-induced increase in expression of TNF-α and IL-1β was suppressed by MK801, an NMDAR antagonist [19]. Hence, it was hypothesized that NMDAR in the WM might be a novel therapeutic target for hypoxic WMD [160].

Microglia express functional mGluRs [161]. The mGluRs include group I (mGluR1, mGluR5), group II (mGluR2, mGluR3) and group III (mGluR4, mGluR6, mGluR7, mGluR8 ) receptors. It has been reported that activation of specific subtypes of mGluRs confers a neuroprotective (group III) or neurotoxic (*via* group II mGluRs) property to microglia [162-165]. The direct activation of group II mGluRs, particularly mGluR2, induces microglial stress such as mitochondrial depolarization and apoptosis [165, 166]. The toxicity of microglial mGluR2 stimulation involves the release of TNF-α and Fas ligand [162], which trigger oligodendroglial caspase-3 activation *via* TNF-R1 and Fas receptor, leading to their death [46]. However, the expression profiles of mGluR in WM microglia under hypoxic conditions have not been thoroughly investigated. Yet, the available data suggest that pathways resulting in group II mGluR stimulation in microglia can be targeted to control microglia-mediated damage to WM.

## Iron Accumulation

Iron, which is abundant in the developing brain [167], is essential for the normal production and maintenance of myelin [168], neurotransmitter release and energy metabolism in the neonatal period [169]. In developing WM, iron was first localised in the microglia. With progression in development, appearance of iron was localized subsequently in oligodendrocytes [170, 171]. Consistent with this, in the developing white matter the expression of iron regulatory proteins and transferrin receptor, which is involved in iron acquisition was localised to the microglia [20, 172]. With progressive activation of microglia there seems to be an increase in iron accumulation in microglia in the vicinity of lesion in several neuropathologies [173-175]. Conditions such as hypoxic-ischemic injury have been reported to increase iron levels in the developing brain [176]. Excess iron promotes oxidative damage and has been implicated in the pathogenesis of neurodegenerative diseases [177]. In conditions such as multiple sclerosis and Alzheimer's disease, the WM tracts were predominantly occupied by iron-laden microglia [178]. In animal models of hypoxia-ischemia, when the oligodendrocyte development is compromised, microglial cells accumulate more iron [179]. This was suggested to be an effort by microglial cells to sequester the excess iron so as to prevent oxidative damage [172]. Under hypoxic conditions iron mediates the release of pro-inflammatory cytokines TNF-α and IL-1β [20]. Also, supplementation with iron was shown to augment the release of TNF-α and IL-1β by microglia when treated with LPS [180]. These findings suggest that iron plays a key role in mediating cytokine production by microglia. Further changes in intracellular iron levels were reported to influence NO production in BV2 microglial cells [181]. In addition, iron-mediated release of free radicals by microglia was shown to induce damage to developing oligodendrocytes in hypoxic WM [20].

## Nitric Oxide Production

NO is produced from L-arginine and the conversion is catalysed by the enzyme nitric oxide synthase (NOS). Based on the source and mode of expression, NOS may be neuronal NOS (nNOS) and endothelial NOS (eNOS) or inducible NOS (iNOS). NO has been implicated in the pathogenesis of hypoxic WMD [182]. Expression of all three isoforms of NOS (iNOS, eNOS and nNOS) was reported in the neonatal brain following hypoxia-ischemia [183-185]. However, of the three isoforms calcium-independent iNOS contributed significantly to the pathophysiological response of the brain to hypoxia [186, 187]. In postnatal brain, iNOS is not constitutively expressed but is expressed by activated microglia under hypoxic conditions [19]. It was shown that hypoxia-induced iNOS expression in microglia is regulated by the phosphatidylinositol 3 kinases (PI3k)/Akt/mammalian target of rapamycin signaling pathway and activation of hypoxia inducible factor-1α [188]. Another study showed that in hypoxic WM, NO production by activated microglia *via* iNOS was mediated by the NF-kB signaling pathway [19]. Enhanced production of NO has been implicated in damaging the myelin producing oligodendrocytes in hypoxic WM and, hence in delayed myelination [19, 30, 189]. Inhibition of iNOS has been reported to decrease edema and apoptotic cell-death and is thought to be neuroprotective in perinatal hypoxia-ischemia [190]. In addition, it has been shown that intraperitoneal injection of 1400W (N-(3-(Aminomethyl)benzyl)acetamidine), a selective iNOS inhibitor, caused a reduction in hypoxia-induced death of oligodendendrocytes in the WM ameliorating WMD [19]. Further, NO-induced mitochondrial dysfunction has been suggested to be associated with oligodendrocyte cell death [189].

## Free Radical Production

Free radicals are derivatives of oxygen (reactive oxygen species (ROS)) and/or nitrogen (reactive nitrogen species (RNS)), and include hydroxyl radical (OH), hydrogen peroxide ($H_2O_2$), singlet oxygen ($1O2$), hypochlorus acid (HOCl), peroxynitrite anion (ONNOO-) and peroxynitrous acid (ONOOH). Free radicals mediate cellular damage including lipid peroxidation, protein modification, and DNA strand breaks. Cellular accumulation of free radicals has been reported in many neurological diseases especially following hypoxic-ischemic insults [15, 191]. Evidence suggests that ROS generation may be implicated in WM lesions following hypoxic-ischemic injury in premature brain [192]. Enhanced production of ROS in the WM was reported in neonatal rats subjected to hypoxia [191]. Activated microglial cells produce superoxide by the enzymatic activity of phagocytic nicotinamide adenine dinucleotide phosphate oxidase (NOX), which releases superoxide in the extracellular space. Active NOX is a multisubunit protein complex that reduces molecular oxygen to superoxide. Expression of mRNAs of all NOX subunits and presence of most of the NOX proteins has been reported in the microglial cells [193-197]. Essential for the superoxide production by Nox are especially the subunits gp91phox and p47phox. Microglial cultures derived from mice deficient in one of these two subunits are unable to produce superoxide following activation [195, 198].

Increased expression of ROS/RNS has been often associated with excess cytokine production [199-203]. It has been shown that cytokines such as IL-3 and granulocyte macrophage colony stimulating factor induced both rapid transient as well as chronic increase in intracellular ROS levels [204, 205]. Another study showed an increase in ROS following administration of IL-1β in conscious rats [206], which was partially blocked by addition of superoxide dismutase and catalase [207]. ROS has been suggested to compromise the integrity of the BBB [208-210]. Along with pro-inflammatory cytokines, hypoxia results in enhanced production of free radicals which have been reported to play a role in the pathogenesis of periventricular WMD [211]. The brain of newborns has been reported to be at a greater risk of free radical-mediated injury as there is evidence of an imbalance between antioxidant and oxidant-generating systems [192, 212]. Antioxidant enzymes such as glutathione have been shown to decrease in the brain following hypoxia-ischemia in neonatal rats [213]. Many studies have demonstrated free radical production in the developing WM in hypoxic-ischemic conditions [192] and microglia have been implicated in their generation [214]. It was shown that ROS production by microglia in hypoxic WM was enhanced following hypoxia and was prevented by edaravone [191], an antioxidant drug that was reported to reduce ischemia-induced inflammation in the brain [215]. It has been speculated that the selective susceptibility of oligodendrocyte progenitors in the WM to hypoxia-induced microglial ROS may be due to their lack of antioxidant enzymes such as glutathione peroxidase and catalase during development [155].

## Proteases Secretion

Proteases, also referred to as peptidase/proteinase, are enzymes that perform hydrolysis of the peptide bonds that link amino acids together in the polypeptide chain forming the protein. In addition to cytokines, hypoxia-induced activation of microglia results in the synthesis and secretion of proteases that are potentially involved in many functions. The proteases synthesized by microglia include; cathepsins B, L, and S, the matrix metalloproteinases (MMP)-1, MMP-2, MMP-3, and MMP-9, and the metalloprotease-disintegrin ADAM8 plasminogen [216]. Following activation by LPS, microglia were demonstrated to produce proteolytic enzymes like MMP-9 [217]. It was shown that MMPs caused damage to WM regions under oxygen-deprived conditions such as stroke [218]. MMPs were also reported to possess the potential to cause damage to the oligodendrocytes [218]. Enhanced expression of cathepsin B, MMP-1, MMP-2 and MMP-3 has been observed in several models of WMD [219-221] in which microglial activation is implicated. Moreover, knockout of MMP-9 was found to reduce damage to WM in the immature brain following hypoxia [222]. Other proteases, such as microglia-derived elastase are thought to influence interaction between neurons and microglia apart from contributing to the breakdown of extracellular matrix [223, 224].

## DRUGS THAT TARGET MICROGLIAL ACTIVATION

The consequences of microglial activation are numerous (as discussed above), and are known to augment hypoxia-induced damage to the developing WM. Apart from neonatal hypoxic WMD, microglial activation has been implicated in a number of other clinical conditions including neurodegenerative diseases, stroke, trauma, demyelinating disorders and neuropathic pain. Though several drugs that suppress microglial activation have been identified, these drugs are far from being used in the treatment of WMD in neonates. However, in this subsection, treatments which may modulate microglial response and thereby reduce hypoxic damage to developing WM are discussed.

## Glutamate Receptor Antagonists

### NMDAR Antagonists

It has been discussed previously that hypoxia-induced activation of NMDAR in microglia results in death of oligodendrocytes in the developing WM [19]. Several NMDAR antagonists have proven to be effective in reducing the effect of microglial response in inflammatory conditions.

### <u>MK801</u>

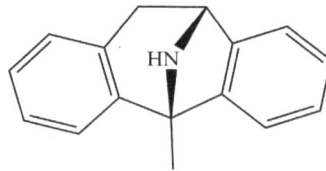

MK801

MK801, also known as Dizocilpine, is a non-competitive antagonist of NMDAR. MK801 blocks NMDAR by binding to a site within the ion channels preventing the entry and exit of calcium across them [225]. MK801 has been shown to block microglial activation and subsequently prevent brain injury [226]. Administration of MK801 reduced the hypoxia-augmented release of TNF-α and IL-1β by microglia in the WM [19]. The same study showed a reduction in death of oligodendrocytes treated with conditioned medium from activated microglia treated with MK801 [19]. It was suggested that NMDAR mediated increase in NO *via* the NF-kB signaling pathway was prevented following MK801 treatment. Moreover, MK801 suppressed the microglial response in the rat brain following forebrain ischemia [227]. In addition, WMD was reduced by pre-treatment with MK-801 in neonatal mice with shaken baby syndrome [228]. Despite repeated success in animal models, MK801 has failed in clinical trials because of its ability to block the physiological NMDAR activity. However, due to its high efficacy, MK801 is widely used as an experimental drug in scientific research to study the effects of NMDAR activation in microglia [19, 226, 229, 230].

## Ketamine

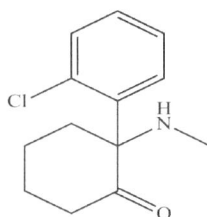

Ketamine

Ketamine is a an anesthetic agent, and is generally reserved for patients with severe hypotension or respiratory depression [231]. Pharmacologically, ketamine is a non-competitive antagonist of NMDAR, that is, it inhibits the receptor by binding to its allosteric site. In addition to its anesthetic property, ketamine has been reported to exert anti-inflammatory effects on macrophages and leukocytes both *in vitro* and *in vivo* studies [232-234]. Shibakawa and his colleagues showed that ketamine (300 ~ 1000 $\mu$M) significantly inhibited inflammatory responses of activated microglia [235]. Another study indicated that ketamine's anti-inflammatory activity may be mediated, in part, by inhibition of extracellular signal regulated kinases (ERK1/2) phosphorylation in primary cultured microglia [236]. Currently, ketamine is used as a paediatric anesthetic agent which provides a unique unconscious state, enabling safe and effective treatment of uncooperative children [237-239]. Because of its safe use in children, we believe that it might be a viable drug for the treatment of hypoxic WMD in neonates through its anti-inflammatory actions.

## Mememtine

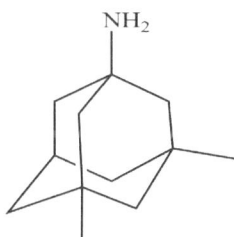

Mementine

Unlike the antagonists of NMDAR described above, memantine's mechanism of antagonism is voltage-dependent, low-moderate affinity and an uncompetitive antagonism with fast-blocking/unblocking kinetics [240]. The low affinity and rapid off-rate kinetics of memantine makes it a favorable NMDAR antagonist, since it preserves the physiological function of the receptor [241-243]. Memantine was identified to inhibit prolonged microglial activation and hence considered to be neuroprotective [244-246]. Also, memantine was shown to attenuate WM injury in rat model of periventricular leukomalacia [247]. Administration of memantine markedly increased the abilities of spatial discrimination, learning and memory, motor coordination, promoted weight gain, and improved long-term prognosis in neonatal rats with periventricular leukomalacia [248]. Currently, the drug has been clinically approved for the treatment of Alzheimer's disease [249, 250] and the use of this drug for the treatment of neonatal WMD is under serious consideration.

### Non-NMDAR Antagonists

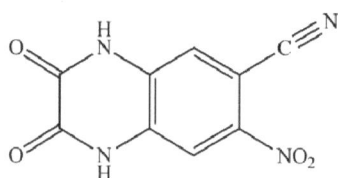

CNQX

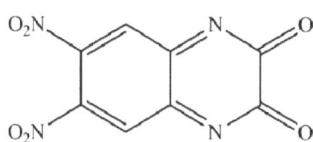

DNQX

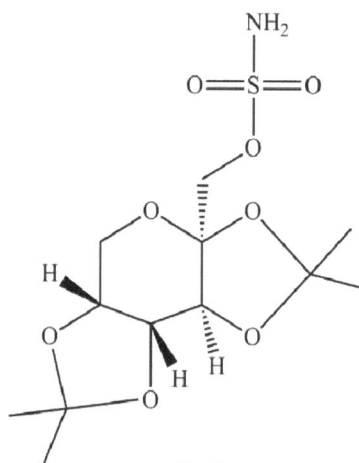

Topiramate

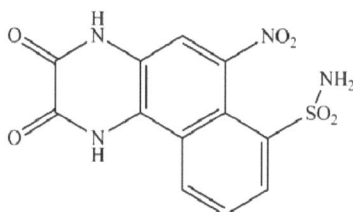

NBQX

Non-NMDAR like AMPARs and KAR play a key role in mediating excitotoxic damage to the developing WM following hypoxia. DNQX

(6,7-dinitro-quinoxaline-2,3-dione), NBQX (2,3-dihydroxy-6-nitro-7-sulfamoyl-benzo[f]quinoxaline-2,3-dione) and CNQX (6-cyano-7-nitroquinoxaline-2,3-dione) are potent and selective antagonists of AMPAR/KAR in the mammalian central nervous system but do not influence NMDAR-mediated responses [251, 252]. Application of AMPAR antagonists such as DNQX and NBQX have been shown to prevent microglial activation [253, 254]. In addition, it was shown that NBQX attenuated damage to WM following hypoxic insults in neonatal rats [255]. Moreover, NBQX and CNQX, blocked the microglial IL-$\beta$-mediated toxicity to immature oligodendrocytes [256]. Topiramate, another AMPAR antagonist [257], was demonstrated to prevent toxicity to oligodendrocytes in periventricular leukomalacia. Also, WM lesions induced in newborn mice were shown to be reduced in animals treated with topiramate [258]. Currently, topiramate is prescribed as an antiepileptic drug in paediatric cases and has proven to be effective [259]. Hence, experimental evidence supports for the consideration of topiramate as a candidate in therapy for perinatal WMD.

## Iron Chelators

Free iron available for chelation is a major catalyst contributing to the initiation and propagation of free radical mediated cascade and is implicated in tissue injury [260, 261]. Activated microglia are often found to accumulate iron [173-175]. In various neurodegenerative models chelators such as deferoxamine, deferiprone and deferasirox have been shown to exert neuroprotection [262-265]. Apart from these chelators, there are natural and synthetic iron chelators which have been proven to have a beneficial effect.

### Deferoxamine

Deferoxamine was the first iron chelator to be approved for clinical use and is derived from the actinobacteria, *Streptomyces pilosus*. Deferoxamine mesylate, apart from its iron chelating property also has limited antioxidant defences [266]. Administration of deferoxamine immediately following hypoxic ischemic injury in postnatal rats, showed a reduction in brain edema [267]. Beneficial effects of treatment with deferoxamine include reduced lipid peroxidation, improved post-ischemic vasoreactivity, cerebral perfusion and adenosine triphosphate (ATP) recovery [268-271]. Deferoxamine preserved the electrocortical activity and cortical cell membrane stability during the early phase of reperfusion in lambs after hypoxic-ischemic insult [272]. Furthermore, deferoxamine has been proven to reduce the consequences of microglial activation. Administration of deferoxamine reduced the number of activated microglia in the intracerebral hemorrhage induced brain injury model [273]. In our recent study, we reported that in primary microglial cells, hypoxia induced up-regulation of TNF-α and IL-1β was reverted to control levels on treatment with deferoxamine [20]. The expression of NF-κB involved in cytokine production cascade [274] was reduced to base line levels in LPS activated microglia on treatment with deferoxamine [180]. In addition, treatment of hypoxic microglial cells with deferoxamine resulted in the reduction of ROS/RNS generation to control levels [20]. Also the toxic effects of conditioned medium from hypoxic microglia on oligodendrocytes were diminished when treated with deferoxamine [20], suggesting it as a possible therapeutic strategy to treat hypoxic/ischemic injury [275, 276].

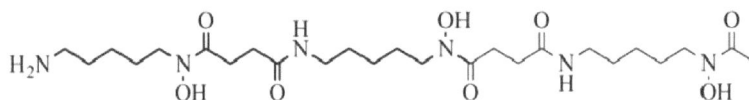

Deferoxamine

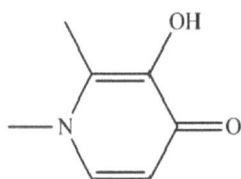

Deferiprone

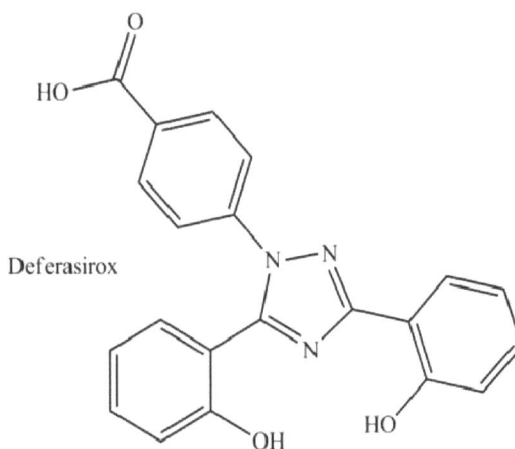

Deferasirox

## Deferiprone and Deferasirox

Deferiprone    (3-hydroxy-1,2-dimethylpyridin-4(1*H*)-one)    and    deferasirox (4-[(3*Z*,5*E*)-3,5-bis(6-oxo-1-cyclohexa-2,4-dienylidene)-1,2,4-triazolidin-1-yl] benzoic acid) were formulated as alternative drugs to deferoxamine with oral efficacy and accessibility across the BBB [277]. In mouse models of autoimmune encephalomyelitis, oral administration of deferiprone reduced the levels of inflammatory cell infiltrates and inhibited T-cell function [278]. In light induced retinal degeneration models deferiprone has been shown to inhibit microglial activation and prevent their migration to the outer layers of retina [279]. Systemic administration of deferiprone and deferasirox in animal models of Parkinson's disease attenuated loss of dopamine neurons and striatal dopamine content [264].

## Other Iron Chelators

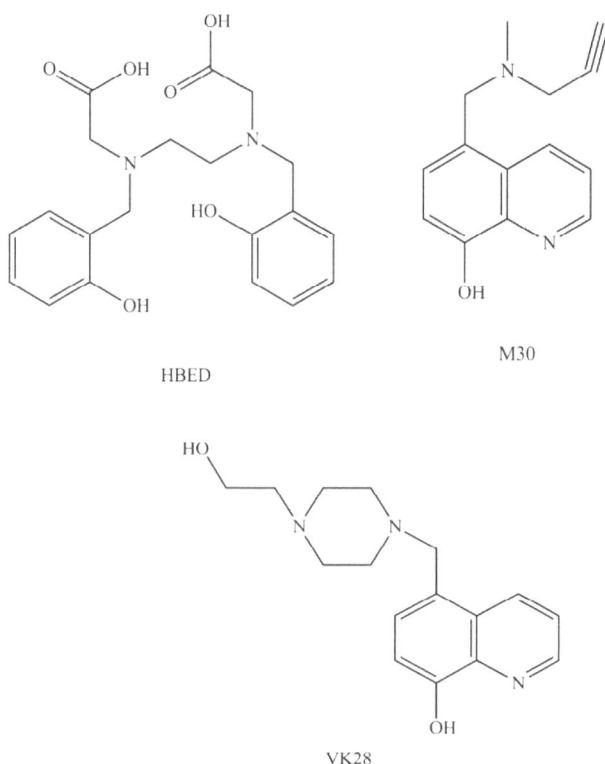

HBED

M30

VK28

Though the above mentioned iron chelators are clinically used, there are certain limitations for their use. Hence the search for non-toxic lipophilic brain-permeable iron chelators led to the synthesis of compounds such as VK-28 (5-(4-(2-hydroxyethyl) piperazin-1-yl (methyl)-8-hydroxyquinoline), M30 (5-(N-methyl-N-propargyaminomethyl)-8-hydroxyquinoline) and HBED (*N,N*_-bis (2-hydroxybenzyl) ethylenediamine-*N,N*_-diacetic acid ). In comparison with deferoxamine, HBED chelates at least twice the amount of iron and its potential in chelating iron has been evaluated in human volunteers and primates [280, 281]. HBED is known to protect the mitochondria by chelating the iron in mitochondria [282]. In an animal model of status epilepticus, HBED attenuated neuronal injury by chelating mitochondrial iron [283-285]. Drugs VK-28 and M-30 possess the propargylamine moiety which could inhibit monoamine oxidase. The *N*-propargylamine is known to up-regulate anti-apoptotic Bcl-2 family proteins (Bcl-2, Bcl-xl) and down-regulate Bad and Bax, thereby contributing to their neuroprotective activity. This is thought to be dependent on the activation of protein kinase C (PKC) and MEK pathways, since inhibitors of these pathways prevented the neuroprotective activity of propargylamines [284-286]. In animal models of Parkinson's disease administration of VK-28 alleviated lipid peroxidation and it was suggested that the neuroprotection conferred by VK-28 is comparable to that of vitamin E [287]. The iron chelator M-30 exhibited neuroprotection in MPTP-induced Parkinsonism [288]. Along with these the naturally available green tea catechins have also been demonstrated to have iron chelating properties. Though the mechanism of action of these drugs in microglia has not been studied, the neuroprotection conferred by these drugs could be due to chelation of excess iron accumulated in microglia in neurodegenerative diseases.

## Imuno-Suppressants and Anti-Inflammatory Drugs

## Glucocorticoids

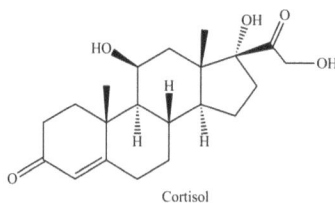

Cortisol

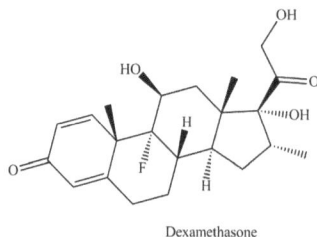

Dexamethasone

Glucocorticoids (GCs) are a group of steroid hormones, synthesized in the adrenal cortex, that exert their action by binding to the glucocorticoid receptor. GCs are potent anti-inflammatory and immunosuppressive drugs. GCs are capable of suppressing the number and the phagocytic activity of macrophages [289]. Subcutaneous injections of GCs in the postnatal rats resulted in drastic reduction in the number of AMCs in the corpus callosum and this was attributed to the reduction in the number of circulating monocytes following GC administration [290, 291]. Furthermore, GCs induced premature differentiation of AMCs to ramified microglial cells [291]. GCs anti-inflamamtory property is demonstrated by suppressing the production of a number of cytokines including IL-12p40 [292]. This could be mediated by attenuation of JNK MAP kinase phosphorylation, as well as by suppressing activator protein (AP)-1 and NF-κB activity [292, 293].

Dexamethasone is a synthetic member of glucocorticoid class of steroid drugs and is widely used for treatment of inflammation and autoimmune disorders [294]. Dexamethasone is known to inhibit microglial activation by attenuating NOX-2 and iNOS [295]. MCP-1, which favors migration of microglia to the site of injury was suppressed by dexamethsone treatment [296]. This was enabled by the ability of dexamethasone to induce MAPK phosphatase (MKP)-1 and thereby suppress p38 and JNK MAP kinase pathways. Administration of dexamethasone in animal models of LPS-induced WM injury, suppressed NOX-2, which is involved in free radical production. In addition, the MKP-1 up-regulation in activated microglia suggests the attenuation of MAPK pathway, involved in inflammation [295].

## Minocycline

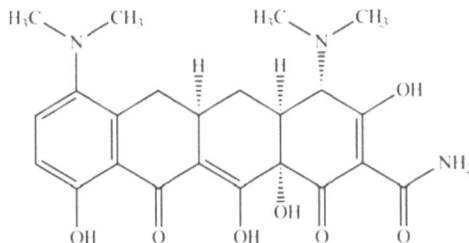

Minocycline

Minocycline is a semi-synthetic tetracycline with anti-inflammatory properties. In many models of neurodegenerative diseases, minocycline has been demonstrated to be neuroprotective [297-299]. Minocycline can effectively inhibit the mitochondrial permeability transition pore and cytochrome c release from mitochondria [297, 300]. In animal models of cerebral hypoperfusion chronic minocycline administration minimized the WMD and the mechanism of action was suggested to be inhibition of microglial activation [301]. In accordance with this, hypoxia-induced up-regulation of toxic factors such as NO and pro-inflammatory cytokines in microglial cells was attenuated by minocycline [302]. The phosphorylation of p38 MAP kinase in activated microglia, which is involved in the inflammatory pathway, was also inhibited by minocycline [302, 303]. In hypoxic-ischemic animal models post-insult administration of minocycline not only reduced the numbers of microglia but also attenuated the WMD [304]. Recently, minocycline was shown to act as an iron chelator attenuating iron overload during brain injury [305].

## Chloroquine

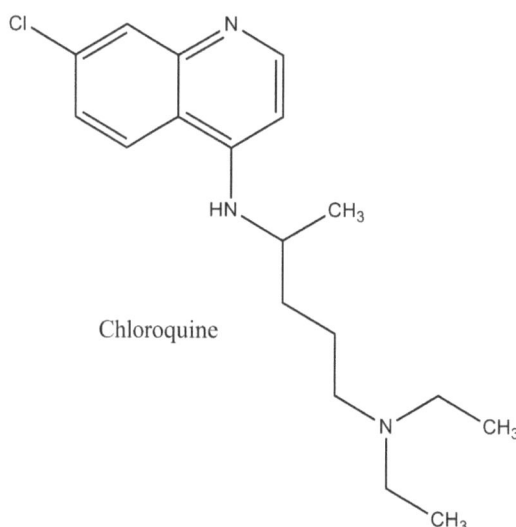

Chloroquine

Chloroquine, is an antimalarial drug with proven anti-inflammatory properties [306]. It exerts its anti-inflammatory effect by down-regulating synthesis of pro-inflammatory cytokines such as TNF-α and IL-1β [307]. Chloroquine attenuates microglial activation by down regulating expression of LT-β, TNF-α,

IL-1α, IL-1β and IL-6 [308]. In rats injected with chloroquine, microglia appeared to have many vacuoles and lysosomes. In addition, chloroquine resulted in internalisation of the plasma membrane to the vacuoles and lysosomes [309]. Treatment with chloroquine did not affect either the phagocytic activity or the antigen-presenting function of microglia, as there was an increase in the expression of complement receptor 3 (CR3) receptors and MHC I antigens [309]. In cultured microglial cells, Aβ accumulation was increased when chloroquine was added to the culture medium [310].

## Non-Steroidal Anti-Inflammatory Drugs

In subsection Phagocytosis, the potential contributions of microglial inflammatory mediators in the pathological progression of neonatal hypoxic WMD were discussed. Based on the evidence, several anti-inflammatory drugs have been tested for the treatment of WMD in the developing brain. Ibuprofen, a non-steroidal anti-inflammatory drug (NSAID), reduced microglial activation [311, 312] and attenuated WMD following hypoxia-ischemia in the immature rodent brain [313]. Other NSAIDs such as aspirin, indomethacin, paracetamol, or NS-398 were tested for their neuroprotective effects in neonatal mice following excitotoxic brain injury and it was shown that pretreatment with low-or high-dose aspirin, and post-treatment with paracetamol or NS-398 protected the brain against WM lesions [314].

Although, these anti-inflammatory drugs have been widely used to suppress microglial activation, anti-inflammatory agents of natural origin are preferred for treatment and are discussed in detail in subsection Natural Extracts.

## Antioxidants

Oxidative stress is a key factor mediating hypoxic-ischemic brain injury in preterm infants. Oxidative stress occurs when there is an imbalance between free radical generation and the antioxidant defence mechanism. The developing brain is highly vulnerable to hypoxic-ischemic injury due to the lack of antioxidant defence mechanisms. Antioxidants are molecules that are capable of protecting the cells from free radical mediated injury. Antioxidants such as melatonin and edaravone have been proven to confer protection against the by-stander injury caused by microglial activation.

## Melatonin

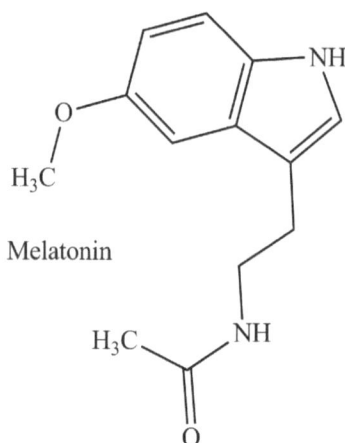

Melatonin

Melatonin (5-methoxy-N-acetyltryptamine) a derivative of amino acid tryptophan, is a neurohormone secreted by the pineal gland, and plays a role in maintaining circardian rhythms and has immunomodulatory actions [315, 316]. Administration of melatonin in different models of WM damage was proven to be beneficial, as melatonin suppressed astrocytosis and microglial activation; and promoted oligodendroglial maturation and myelination repair [199, 317-319]. Furthermore, melatonin enhanced axonal re-growth and sprouting [320]. Melatonin increased the cell numbers of AMCs and also induced the expression of CR3, MHC I and MHC II, and CD4 antigens [321], leading to enhanced endocytic and antigen-presenting capacity. This increased immune potentiality and its maintenance in microglia required continuous action of the drug. Melatonin is well known to exhibit anti-inflammatory action by inhibiting inflammatory cytokines production in microglia [322]. Treatment with melatonin resulted in reduced chemokine expression in LPS induced BV2 murine microglial cells through inhibition of NF-κB [322]. Melatonin inhibited amphetamine-induced iNOS mRNA overexpression in microglial cell lines [323]. Besides its role in modulating immune function of microglia, melatonin is also known to suppress the free radical generation. Recent studies have shown that melatonin decreases superoxide production in activated microglia by impairing nicotinamide adenine dinucleotide phosphate (NADPH) oxidase assembly [324].

## Edaravone

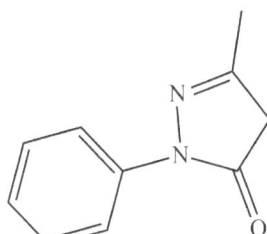

Edaravone

Edaravone (3-methyl-1-phenyl-2-pyrazoline-5-one/MCI-186), a neuroprotective agent, is a free radical scavenger known to reduce the extent of damage in both grey and WM regions following brain injury [325]. It exerts its action by binding to the hydroxyl radicals and inhibits hydroxyl radical mediated lipid peroxidation [326]. This neuroprotective agent has been documented to reduce brain edema [327, 328], inhibit vascular endothelial injury [326], extend the therapeutic time window [329] and improve neurological deficits [326]. Edaravone is known to scavenge hydroxyl radicals and peroxynitrite produced by activated microglia [330] and can suppress accumulation of 4 hydroxy-2 nonenal modified protein following an ischemic injury [215]. In animal models of transisent focal ischemia, along with its antioxidant property, edaravone was also found to exert its anti-inflamamtory property by reducing the number of activated microglia and by inhibiting iNOS expression in microglia [215]. On treating hypoxic microglial cells with edaravone there was a significant reduction in the release of pro-inflammatory cytokines such as TNF-α and IL-1β and chemokines such as CCL2 and CXCL12 [191]. This effect of edaravone on hypoxic microglial cells was proposed to be a therapeutic strategy for treating WMD [191].

## Ebselen

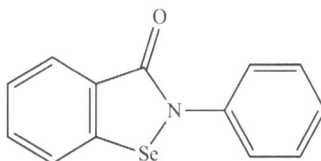

Ebselen

Ebselen was developed as an anti-inflammatory agent but was later established as an antioxidant due to its ability to mimic glutathione peroxidase, an antioxidant enzyme [331]. Ebselen is reported to limit free radical mediated damage by inhibiting 5-and 15-lipoxygenases [332], iNOS [333], NADPH oxidase [334] and also by deactivating peroxinitrite [335]. In addition, ebselen exhibits its anti-inflammatory property by down regulating NF-κB translocation and *jun*-N-terminal kinase phosphorylation in the inflammatory cytokines production cascade [336]. Microglia play a key role in inflammation mediated brain damage in ischemia [337]. Hence it was suggested that the protective effect of ebselen could be due to inhibition of microglial activation [338]. Post ischemic administration of ebselen was proven to be beneficial as it reduced oligodendrocyte damage and inhibited lipid peroxidation [339]. Further, ebselen reduced the number of amyloid precursor protein (APP) accumulating axons and Tau positive oligodendrocytes, and thus attenuated WM damage [339].

## Vitamins

Retinoic acid

Vitamin E

Evidence suggests that vitamins have a role in brain functions and posses anti-inflammatory property. Vitamin E (α-tocopherol) is a biological antioxidant and is known to suppress microglial activation [340, 341]. On treatment with vitamin E, the phosphorylation of p38 MAPK, activation of NF-κB and subsequent expression of iNOS was attenuated in microglial cells activated with LPS [341]. Consistent with this, when LPS activated BV2 cells were supplemented with vitamin E, phosphatase 2A activity was increased, and cyclooxygenase (COX)-2 transcription was inhibited. This was mediated by silencing PKC/ERK/NF-κB-signaling cascade [342]. In cultured microglial cells, vitamin E induced transformation of AMC to RMC and this was accompanied by down-regulation of adhesion molecules leukocyte function antigen-1, very late antigen-4, and intercellular adhesion molecule-1. This shift in morphology was suggested to immunologically de-activate the microglial cells [343].

All-trans retinoic acid (RA) is a vitamin A metabolite and is known to have a role in cellular differentiation, CNS development and homeostasis [344]. Evidence suggests that RA has anti-inflammatory properties. It attenuates the production of inflammatory molecules such as TNF-α and iNOS [345, 346]. In LPS/β-amyloid-induced activation of microglial cells, RA inhibited the expression of TNF-α and NO [347, 348]. It was shown that RA enhanced the transcription of TGF-β1 and RA receptor (RAR) β1. Further, it blocked the translocation of NF-κB in the activated microglial cells [348]. In LPS activated microglia, on concomitant treatment with RA, the expression of CR3 on microglia was suppressed [104]. In view of these results, RA was suggested as a potential therapeutic agent owing to its anti-inflammatory response and efficacy in inhibiting activation of microglia.

## Ion Channels Blockers

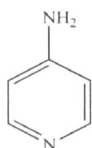

4-aminopyridine

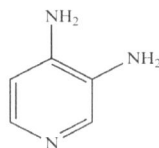

3,4 diaminopyridine

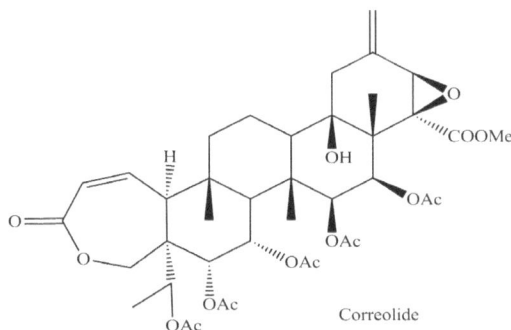

Correolide

A growing body of evidence points to ion channels on microglia as major contributors of WM injuries. Patch-clamp studies of microglial cells in cell culture and in tissue slices demonstrated that microglia express a wide variety of ion channels. Six different types of K+ channels have been identified in microglia, namely, inward rectifier (Kir2.1), delayed rectifier (Kv1.3, Kv1.5), HERG-like (Kv11.1-like), G protein-activated, as well as voltage-dependent and voltage-independent $Ca^{2+}$-activated K+ channels [349, 350]. Moreover, microglia express H+ channels, Na+ channels, voltage-gated $Ca^{2+}$ channels, $Ca^{2+}$-release activated $Ca^{2+}$ channels, and voltage-dependent and voltage-independent Cl−channels [349, 350]. Expression patterns of ion channels in microglia depend on the microglial activation state. Microglial ion channels can be modulated by exposure to activating stimuli such as LPS, cytokines, by activation of PKC or G proteins, by factors released from astrocytes, by changes in the concentration of internal free $Ca^{2+}$, and by variations of the internal or external pH [21].

In light of the above, voltage-gated potassium (Kv) blockers have been considered for the treatment of WMD in which microglial activation is implicated. Non-specific blockers of Kv channels, such as 4-amino-pyridine and 3,4-diaminopyridine, have been tested clinically for their efficacy in the treatment of WM disorders in ageing population. Also highly selective Kv channel blockers such as margatoxin, correolide, kaliotoxin, ShK, and Sh-Dap22 have been used in experimental models of WM injury and are being considered for clinical trials [351-354]. But their efficacy in infants still remains to be investigated. Given the increased expression of Kv1.2 in activated microglia under hypoxic conditions and its involvement in enhancing the expression of IL-1β and TNF-α [355], it is believed that microglial Kv channels might be a potential target for the treatment of neonatal WMD.

Calcium-activated-K+ channels (KCNN4/KCa3.1/SK4/IK1) are highly expressed in rat microglia [356, 357] and are potential therapeutic target for WM injuries, since they may directly affect inflammation. $Ca^{2+}$ signaling plays an important role in microglial response to activation including proliferation, migration, phagocytosis, production of pro-inflammatory cytokines/chemokines and ROS [358-361]. Following activation, the increase in intracellular $Ca^{2+}$ activated $Ca^{2+}$ dependant K+ channels [362-364]. $Ca^{2+}$/calmodulin-activated K+ channels, in particular KCa3.1 (KCNN4/IK1/SK4) are the subject of many investigations from a drug development point of view. KCa3.1-selective blocker triarylmethane-34 (TRAM-34) [365] was used to assess whether KCa3.1 channels contributed to microglial activation. It was shown that KCa3.1 channels contribute to microglia activation, iNOS up-regulation, production of NO and peroxynitrite, and to consequent neurotoxicity. Microglia activation involved the signaling pathways-p38 mitogen-activated protein kinase and NF-kB, which are important for up-regulation of numerous pro-inflammatory molecules, and the KCa3.1 channels were functionally linked to activation of p38 but not NF-kB [366]. Hence, microglial calcium-dependant potassium channels are believed to be a potential target for the treatment of hypoxic WMD.

## Adenosine Triphosphate Channel Blockers

ATP released in the extracellular space serves as a signaling molecule mediating interaction between various cell types in the brain by binding to the purinoceptors expressed on these cells. The purinoceptors are broadly classified into ionotropic purinoceptors (P2X) and metabotropic purinoceptors (P2Y), of which P2X4, P2X7, P2Y6 and P2Y12 are expressed by microglia [367, 368]. Accumulating evidence suggests the deleterious role of purinergic signaling system in the WMD [367-369]. Free ATP released in excess during hypoxia is implicated in microglial activation [355] and activation of P2X7 and P2X4 receptors on microglia is associated with inflammation [367, 368]. ATP could also induce microglial chemotaxis in response to local CNS injury by activation of P2Y12 receptors [370]. Blockers of purinoceptors such as pyridoxal phosphate 6-azophenyl-2, 40-disulphonic acid (PPADS), suramin, 2', 3'-0-(2, 4, 6-Trinitrophenyl) adenosine 5'-triphosphate (TNP-ATP) have been tested experimentally for their efficacy in attenuating microglial activation. In BV2 microglial cells stimulated with IFN-γ, ATP-induced enhanced production of NO was significantly suppressed by suramin [371]. PPADs and TNP-ATP effectively blocked the secretion of pro-inflammatory cytokines such as TNF-α and IL-1β from microglia subjected to hypoxia [368]. In addition, these drugs resulted in morphological changes in microglia by inducing their ramification [368]. Furthermore, ATP induced PI3K/Akt activation leading to microglial chemotaxis was effectively inhibited by P2X antagonist TNP-ATP [372] by reducing the inflow of calcium through P2X4 receptor. Suramin, but not PPADs was reported to block the ATP induced calcium currents in microglia during their activation [373].

## Natural Extracts

### Ginsenosides

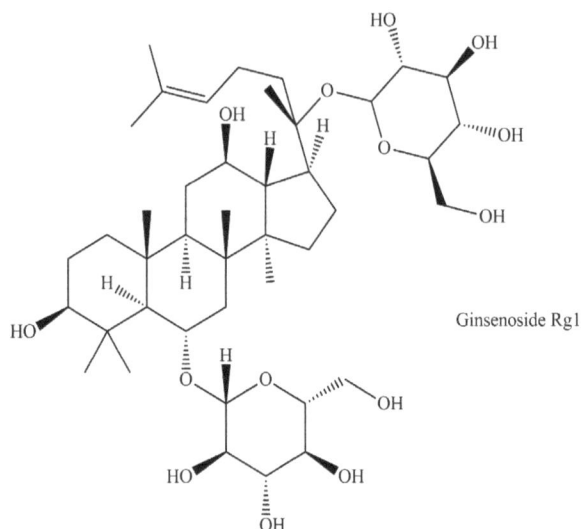

Ginsenoside Rg1

Ginsenosides are a class of steroid glycosides and triterpene saponins found in plants of genus *Panux*. Ginsenosides are known to have anti-inflammatory and immunomodulatory properties [374].

More than 30 ginsenosides have been identified and are classified into two categories: (1) the 20(S)-protopanaxadiol (PPD) (Rb1, Rb2, Rb3, Rc, Rd, Rg3, Rh2, Rs1) and (2) the 20(S)-protopanaxatriol (PPT) (Re, Rf, Rg1, Rg2, Rh1) [375]. All forms of ginsenosides inhibited LPS-induced TNF-α production in N9 microglial cells [376] whereas the NO production in microglia was inhibited by ginsenoside-Rg1 and –Re [376] showing the differential action of different ginsenosides. In LPS/IFNγ activated BV2 microglial cells, ginsenoside Rh2/Rg1 decreased the expression of iNOS, TNF-α, IL-1β and COX-2 [377, 378]. This anti-inflammatory process of Rh2 was reported to be mediated *via* the activator protein (AP)-1 and protein kinase A pathway [378] whereas that of Rg2 was mediated through phospholipase C –γ 1 signaling pathway [377]. Ginsenoside Rg3 up-regulated the expression of macrophage scavenger receptor in primary microglia and enhanced the microglial phagocytosis of Aβ [379]. Though the ability of ginsenosides to alleviate WMD is not yet known, their anti-inflammatory actions may confer protection against damage.

## Gastrodin

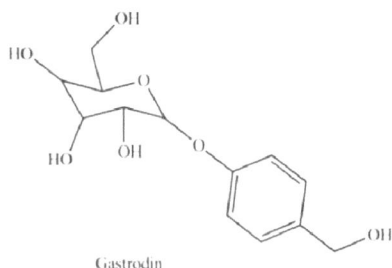

Gastrodin

Gastrodin is the glucoside of 4-hydroxybenzyl alcohol and the active component of *Rhizoma gastrodiae*. The properties of gastrodin include anti-inflammation [380], antioxidation [381] and immunomodulation [382]. Pretreatment with gastrodin significantly increased the survival of cortical neurons following hypoxia by inhibiting the NMDA receptor and glutamate mediated neurotoxicity [380]. Gastrodin reduced the expression of iNOS, COX-2 and pro-inflammatory cytokines in LPS activated microglia. This was achieved through inhibition of phosphorylation of inhibitor of NF-κB kinase subunit beta of the NF-κB pathway and cAMP response element binding (CREB) of MAPK pathway [383]. Furthermore, gastrodin was also found to down regulate JNK MAP kinase, a key molecule involved in neuroinflammation [384, 385]. The anti-oxidative property of gastrodin is exhibited by the reduction in generation of ROS and its anti-apoptotic property is ascribed to the decrease in Bax/Bcl-2 ratio, as well as caspase-3 cleavage and poly (ADP-ribose) polymerase (PARP) proteolysis [386]. In view of the above, gastrodin may be considered as a potential drug for ameliorating WMD.

## Resveratol

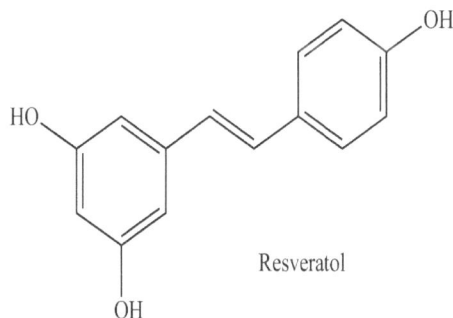

Resveratol

Resveratrol (*trans*-3,4,5-trihydroxystilbene), present in grapes and various medicinal plants, is a phytoalexin. The pharmacological activity of resveratol includes anti-inflammatory and antioxidant effects [387]. Resveratol exhibits anti-inflammatory properties by inhibiting pro-inflammatory mediators such as NF-κB and AP-1 [388]. In primary microglia challenged with LPS, resveratol was able to effectively inhibit free radical production by inhibiting prostaglandin E2 (PGE2) and 8-iso-prostaglandin F2α (8-iso-PGF2α) production [389].

In addition resveratrol reduced the production of pro-inflammatory factors by inhibiting signaling pathways such as MAPKs, PI3-K/Akt, and glycogen synthase kinase-3β (GSK-3β) and thereby inhibited microglial activation [386]. Studies in N9 microglial cells have demonstrated the efficacy of resveratrol in reducing the production of NO and TNF-α by attenuating the LPS-induced phosphorylation of p38 MAPK and degradation of IκB-α [390]. Furthermore, resveratrol rendered its antioxidant property by inhibiting NADPH oxidase and by scavenging free radicals [391]. In addition, administration of resveratol in animal models of hypoxia-ischemia, reduced the extent of WMD by preserving myelination [392].

## Curcumin

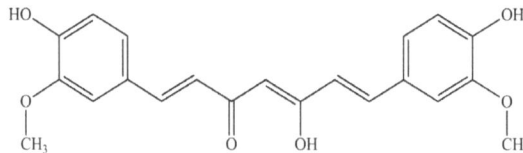

Curcumin

Curcumin ((IE,6E)-1,7-bis(4-hydroxy-3-methoxyphenyl)-1,6-heptadiene-3,5-dione), the yellow pigment of the widely used spice turmeric, is known to have powerful antioxidant and anti-inflammatory properties. The anti-inflammatory action of curcumin was demonstrated by its ability to suppress the production of cytotoxic mediators such as NO and pro-inflamamtory cytokines such as TNF-α, IL-1α and IL-6 and thereby blocking microglial activation [393]. In LPS induced activation of microglial cells, curcumin blocked expression of COX2 by inhibiting transcription factors such as NFkB, AP1, and signal transducers and activators of transcription (STATs) [394, 395]. In addition curcumin induced the expression of anti-inflamamtory molecules such as peroxisome proliferator activated receptor α

(PPARα) and IL4 in microglia [396]. Furthermore, curcumin attenuated microglial migration and up-regulated the expression of genes such as netrin-G1, delta-like-1, platelet endothelial cell adhesion molecule 1, and plasma cell endoplasmic reticulum protein 1, which are involved in cell adhesion and motility [396].

In a co-culture of primary microglia and pre-oligodendrocytes challenged with LPS, curcumin significantly inhibited the apoptosis of pre-oligodendrocytes by attenuating the expression of either gp91phox or p67phox of NADPH oxidase and iNOS [397] in activated microglia which are known to be involved in generation of free radicals. Besides the above, in animal models of LPS induced WMD, curumin administration alleviated the WM from extensive damage by inhibiting loss of pre-oligodendrocytes [397].

## Sesquiterpenes

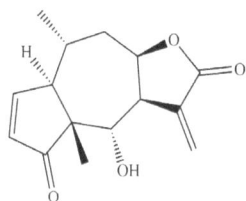 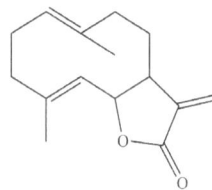

Helenalin                    Costunolide

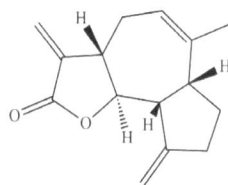

Eremanthin

Sesquiterpenes belong to the family of terpenes and are naturally found in plants and possess anti-inflamamtory properties. Several sesquiterpenes have been identified. By alkylating the p65 subunit of the NF-κB, the sesquinterpene helenalin has been shown to prevent the DNA binding activity of NF-κB [398] resulting in attenuation of inflammatory response. In the brains of diabetic rats, administration of sesquiterpenes such as the costunolide and eremanthine isolated from the plant *Costus speciosus*, significantly increased the levels of glutathione (GSH), accompanied by the increased enzymatic activities of superoxide dismutase (SOD), catalase (CAT) and glutathione peroxidase (GPx) [399]. Sesquiterpenes isolated from the fruits of *Celastrus orbicultus* were able to inhibit NO production in LPS activated BV2 microglia [400]. In addition, a similar response was reported with helenalin [401].

Consistent with this, costunolide effectively attenuated the expression of TNF-α, IL-1,IL-6, iNOS, MCP-1 and COX-2 [402]. This anti-inflammatory action was established by inhibition of NF-κB and MAP kinase pathway [401, 402]. Extracts of *Petasites hybridus* attenuated prostaglandin E2 and COX-2 expression in LPS treated primary microglia by inhibiting phosphorylation of p44/p42 MAPK [403]. Owing to the anti-inflammatory action of different sesquiterpenes, we suggest that they could be considered as alternative therapeutic agents for WMD.

## Colchicine

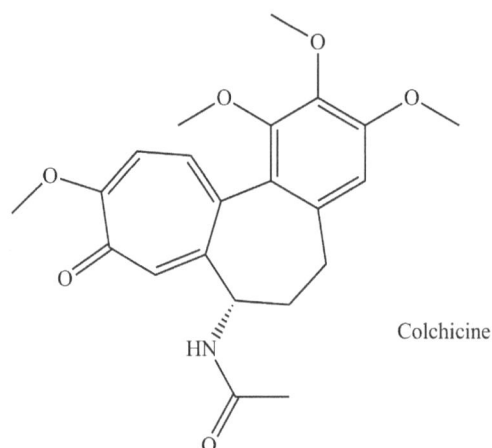

Colchicine

Colchicine is a natural product obtained from plants of *Colchicum*. The drug is known to distrupt the polymerisation of microtubules [404, 405] and induce apoptosis in different cell types. In postnatal rat brains administration of colchicine reduced the number of AMC in the corpus callosum [406] and induced the early differentiation of AMC into the ramified form. In accordance with this, in excitotoxic brain injury models, colchicine when administered in combination with choroquine was able to reduce the numbers of resident and monocytes derived microglia thereby ameliorating the ibotenate induced brain lesions [407].

Although most of the above mentioned drugs were highly effective *in vitro,* their *in vivo* effects were less evident owing to their restricted entry across the BBB. The BBB being a highly regulated barrier, permits selective transportation of molecules across it making the effective delivery of drugs for the treatment of WMD a challenge. Recent progress in understanding the molecular properties of the BBB and advances in medical technology has led to novel models and nanotechnology-based approaches for targeted drug delivery across the BBB [408]. It is important to note that, the BBB in the developing brain is immature [409, 410]. Further, the immature BBB may be weakened by hypoxic-ischemic insults causing increased vascular permeability [411, 412]. Hence, compromised BBB integrity may allow the entry of drugs which normally are unable to traverse a mature functional BBB. In certain cases where the BBB is intact and impermeable to the drugs, application of biological/chemical/physical stimuli is known to temporarily disrupt the BBB allowing concentrated drug to be delivered locally [413-415]. As discussed in the previous sections, several studies have implicated activation of microglia in the pathogenesis of WMD in hypoxic brain injuries and hence it is essential to design drugs that specifically target activated microglia. A recent study demonstrated that neutral nanopolymers localize in activated microglia in the presence of neuroinflammation [416]. This indicates the prospective use of nanoparticles as effective drug delivery vehicles especially in neuroinflammatory conditions, such as WMD [417, 418].

## CONCLUSIONS

Drugs that specifically target microglial activation are not yet available; however, the use of anti-inflammatory agents, antioxidants, iron chelators and pharmacological antagonists of glutamate receptors represent viable therapies for the treatment of hypoxic WMD in which microglial activation is implicated. These approaches could modulate microglial function following hypoxic injury. Given the complex interplay of microglial responses following hypoxia, a promising approach might be to pursue strategies that result in favorable modulation of the genomic and phenotypic state of activated microglia. As described above, several studies on microglia have identified an array of molecular targets and signaling systems modification of which could ameliorate hypoxic damage to developing WM. A comprehensive analysis of the studies on understanding the molecular mechanisms underlying microglial activation in neonates would, hence, be useful to optimize and refine therapeutic approaches for treatment of hypoxic WMD.

## ACKNOWLEDGEMENTS

This work was supported by the research grant (R-181-000-120-213) from National Medical Research Council, Singapore.

## CONFLICT OF INTEREST

There is no conflict of interest among the authors.

## REFERENCES

[1]     World Health Organization; March of Dimes; The Partnership for Maternal NCHStC. World Health Organization; March of Dimes; The Partnership for Maternal, Newborn & Child Health; Save the Children. World Health Organ Tech Rep Ser 2012.

[2]     Johnston MV. Hypoxic and ischemic disorders of infants and children. Lecture for 38th meeting of Japanese Society of Child Neurology, Tokyo, Japan, July 1996. Brain Dev 1997;19(4):235-9.

[3]     McQuillen PS, Ferriero DM. Selective vulnerability in the developing central nervous system. Pediatr Neurol 2004;30(4):227-35.

[4]     Rezaie P, Dean A. Periventricular leukomalacia, inflammation and white matter lesions within the developing nervous system. Neuropathology 2002;22(3):106-32.

[5]     Volpe JJ. Cerebral white matter injury of the premature infant-more common than you think. Pediatrics. 2003 Jul;112(1 Pt 1):176-80.

[6]     Follett PL, Deng W, Dai W, Talos DM, Massillon LJ, Rosenberg PA, *et al*. Glutamate receptor-mediated oligodendrocyte toxicity in periventricular leukomalacia: a protective role for topiramate. J Neurosci. 2004 May 5;24(18):4412-20.

[7]     Folkerth RD. Periventricular leukomalacia: overview and recent findings. Pediatr Dev Pathol. 2006 Jan-Feb;9(1):3-13.

[8]     Vannucci RC, Vannucci SJ. Perinatal hypoxic-ischemic brain damage: evolution of an animal model. Dev Neurosci. 2005 Mar-Aug;27(2-4):81-6.

[9]     Alvarez-Diaz A, Hilario E, de Cerio FG, Valls-i-Soler A, Alvarez-Diaz FJ. Hypoxic-ischemic injury in the immature brain-Key vascular and cellular players. Neonatology. 2007;92(4):227-35.

[10]    Baud O, Daire JL, Dalmaz Y, Fontaine RH, Krueger RC, Sebag G, *et al*. Gestational hypoxia induces white matter damage in neonatal rats: a new model of periventricular leukomalacia. Brain Pathol. 2004 Jan;14(1):1-10.

[11]    Huang BY, Castillo M. Hypoxic-ischemic brain injury: imaging findings from birth to adulthood. Radiographics. 2008 Mar-Apr;28(2):417-39; quiz 617.

[12]    Lyons SA, Kettenmann H. Oligodendrocytes and microglia are selectively vulnerable to combined hypoxia and hypoglycemia injury *in vitro*. J Cereb Blood Flow Metab. 1998 May;18(5):521-30.

[13]  Mitrovic B, Ignarro LJ, Montestruque S, Smoll A, Merrill JE. Nitric-Oxide as a Potential Pathological Mechanism in Demyelination-Its Differential-Effects on Primary Glial-Cells *in vitro*. Neuroscience. 1994 Aug;61(3):575-85.

[14]  Dewar D, Underhill SM, Goldberg MP. Oligodendrocytes and ischemic brain injury. J Cereb Blood Flow Metab. 2003 Mar;23(3):263-74.

[15]  Deng YY, Lu J, Ling EA, Kaur C. Role of microglia in the process of inflammation in the hypoxic developing brain. Front Biosci (Schol Ed). 2011;3:884-900.

[16]  Graeber MB, Li W, Rodriguez ML. Role of microglia in CNS inflammation. FEBS Lett. 2011 Dec 1;585(23):3798-805.

[17]  Kaur C, Ling EA. Periventricular white matter damage in the hypoxic neonatal brain: role of microglial cells. Prog Neurobiol. 2009 Apr;87(4):264-80.

[18]  Baud O, Foix-L'Helias L, Kaminski M, Audibert F, Jarreau PH, Papiernik E, *et al*. Antenatal glucocorticoid treatment and cystic periventricular leukomalacia in very premature infants. N Engl J Med. 1999 Oct 14;341(16):1190-6.

[19]  Murugan M, Sivakumar V, Lu J, Ling EA, Kaur C. Expression of N-methyl D-aspartate receptor subunits in amoeboid microglia mediates production of nitric oxide *via* NF-kappaB signaling pathway and oligodendrocyte cell death in hypoxic postnatal rats. Glia. 2011 Apr;59(4):521-39.

[20]  Rathnasamy G, Ling EA, Kaur C. Iron and iron regulatory proteins in amoeboid microglial cells are linked to oligodendrocyte death in hypoxic neonatal rat periventricular white matter through production of pro-inflammatory cytokines and reactive oxygen/nitrogen species. J Neurosci. 2011 Dec 7;31(49):17982-95.

[21]  Eder C. Regulation of microglial behavior by ion channel activity. J Neurosci Res. 2005 Aug 1;81(3):314-21.

[22]  Villapol S, Fau S, Renolleau S, Biran V, Charriaut-Marlangue C, Baud O. Melatonin Promotes Myelination by Decreasing White Matter Inflammation After Neonatal Stroke. Pediatr Res. 2011 Jan;69(1):51-5.

[23]  Filley CM. White matter and behavioral neurology. White Matter in Cognitive Neuroscience: Advances in Diffusion Tensor Imaging and Its Applications. 2005;1064:162-+.

[24]  Obonai T, Takashima S. In utero brain lesions in SIDS. Pediatr Neurol. 1998 Jul;19(1):23-5.

[25]  Skoff RP, Bessert DA, Barks JD, Song D, Cerghet M, Silverstein FS. Hypoxic-ischemic injury results in acute disruption of myelin gene expression and death of oligodendroglial precursors in neonatal mice. Int J Dev Neurosci. 2001 Apr;19(2):197-208.

[26]  Ness JK, Romanko MJ, Rothstein RP, Wood TL, Levison SW. Perinatal hypoxia-ischemia induces apoptotic and excitotoxic death of periventricular white matter oligodendrocyte progenitors. Dev Neurosci-Basel. 2001 May-Jun;23(3):203-8.

[27]  Kaur C, You Y. Ultrastructure and function of the amoeboid microglial cells in the periventricular white matter in postnatal rat brain following a hypoxic exposure. Neurosci Lett. 2000 Aug 18;290(1):17-20.

[28]  Deng Y, Lu J, Sivakumar V, Ling EA, Kaur C. Amoeboid microglia in the periventricular white matter induce oligodendrocyte damage through expression of pro-inflammatory cytokines *via* MAP kinase signaling pathway in hypoxic neonatal rats. Brain Pathol. 2008 Jul;18(3):387-400.

[29]  Skoff RP, Knapp PE. Lineage and differentiation of oligodendrocytes in the brain. Ann N Y Acad Sci. 1991;633:48-55.

[30]   Kaur C, Sivakumar V, Ang LS, Sundaresan A. Hypoxic damage to the periventricular white matter in neonatal brain: role of vascular endothelial growth factor, nitric oxide and excitotoxicity. J Neurochem. 2006 Aug;98(4):1200-16.

[31]   Gilmore JH, Lin W, Gerig G. Fetal and neonatal brain development. Am J Psychiatry. 2006 Dec;163(12):2046.

[32]   Pang Y, Cai Z, Rhodes PG. Effects of lipopolysaccharide on oligodendrocyte progenitor cells are mediated by astrocytes and microglia. J Neurosci Res. 2000 Nov 15;62(4):510-20.

[33]   Biran V, Joly LM, Heron A, Vernet A, Vega C, Mariani J, *et al.* Glial activation in white matter following ischemia in the neonatal P7 rat brain. Exp Neurol. 2006 May;199(1):103-12.

[34]   Liu Y, Silverstein FS, Skoff R, Barks JD. Hypoxic-ischemic oligodendroglial injury in neonatal rat brain. Pediatr Res. 2002 Jan;51(1):25-33.

[35]   Cai Z, Pang Y, Xiao F, Rhodes PG. Chronic ischemia preferentially causes white matter injury in the neonatal rat brain. Brain Res. 2001 Apr 13;898(1):126-35.

[36]   Jelinski SE, Yager JY, Juurlink BH. Preferential injury of oligodendroblasts by a short hypoxic-ischemic insult. Brain Res. 1999 Jan 2;815(1):150-3.

[37]   Kurumatani T, Kudo T, Ikura Y, Takeda M. White matter changes in the gerbil brain under chronic cerebral hypoperfusion. Stroke. 1998 May;29(5):1058-62.

[38]   Haynes RL, Folkerth RD, Keefe RJ, Sung I, Swzeda LI, Rosenberg PA, *et al.* Nitrosative and oxidative injury to premyelinating oligodendrocytes in periventricular leukomalacia. J Neuropathol Exp Neurol. 2003 May;62(5):441-50.

[39]   Kadhim H, Tabarki B, Verellen G, De Prez C, Rona AM, Sebire G. Inflammatory cytokines in the pathogenesis of periventricular leukomalacia. Neurology. 2001 May 22;56(10):1278-84.

[40]   Deguchi K, Mizuguchi M, Takashima S. Immunohistochemical expression of tumor necrosis factor alpha in neonatal leukomalacia. Pediatr Neurol. 1996 Jan;14(1):13-6.

[41]   Mallard C, Welin AK, Peebles D, Hagberg H, Kjellmer I. White matter injury following systemic endotoxemia or asphyxia in the fetal sheep. Neurochem Res. 2003 Feb;28(2):215-23.

[42]   Kannan S, Saadani-Makki F, Muzik O, Chakraborty P, Mangner TJ, Janisse J, *et al.* Microglial activation in perinatal rabbit brain induced by intrauterine inflammation: detection with 11C-(R)-PK11195 and small-animal PET. J Nucl Med. 2007 Jun;48(6):946-54.

[43]   Billiards SS, Haynes RL, Folkerth RD, Trachtenberg FL, Liu LG, Volpe JJ, *et al.* Development of microglia in the cerebral white matter of the human fetus and infant. J Comp Neurol. 2006 Jul 10;497(2):199-208.

[44]   Hirayama A, Okoshi Y, Hachiya Y, Ozawa Y, Ito M, Kida Y, *et al.* Early immunohistochemical detection of axonal damage and glial activation in extremely immature brains with periventricular leukomalacia. Clin Neuropathol. 2001 Mar-Apr;20(2):87-91.

[45]   Zawadzka M, Kaminska B. Immunosuppressant FK506 affects multiple signaling pathways and modulates gene expression in astrocytes. Mol Cell Neurosci. 2003 Feb;22(2):202-9.

[46]   Deng YY, Lu J, Ling EA, Kaur C. Microglia-derived macrophage colony stimulating factor promotes generation of pro-inflammatory cytokines by astrocytes in the periventricular white matter in the hypoxic neonatal brain. Brain Pathol. 2010 Sep;20(5):909-25.

[47]  Di Giovanni S, Movsesyan V, Ahmed F, Cernak I, Schinelli S, Stoica B, *et al.* Cell cycle inhibition provides neuroprotection and reduces glial proliferation and scar formation after traumatic brain injury. Proc Natl Acad Sci U S A. 2005 Jun 7;102(23):8333-8.

[48]  Di Giovanni S, De Biase A, Yakovlev A, Finn T, Beers J, Hoffman EP, *et al. In vivo* and *in vitro* characterization of novel neuronal plasticity factors identified following spinal cord injury. J Biol Chem. 2005 Jan 21;280(3):2084-91.

[49]  Ment LR, Stewart WB, Fronc R, Seashore C, Mahooti S, Scaramuzzino D, *et al.* Vascular endothelial growth factor mediates reactive angiogenesis in the postnatal developing brain. Brain Res Dev Brain Res. 1997 May 20;100(1):52-61.

[50]  Nakamura Y, Okudera T, Hashimoto T. Vascular architecture in white matter of neonates: its relationship to periventricular leukomalacia. J Neuropathol Exp Neurol. 1994 Nov;53(6):582-9.

[51]  Folkerth RD. Neuropathologic substrate of cerebral palsy. J Child Neurol. 2005 Dec;20(12):940-9.

[52]  Okoshi Y, Itoh M, Takashima S. Characteristic neuropathology and plasticity in periventricular leukomalacia. Pediatr Neurol. 2001 Sep;25(3):221-6.

[53]  Sridhar K, Kumar P, Katariya S, Narang A. Postasphyxial encephalopathy in preterm neonates. Indian J Pediatr. 2001 Dec;68(12):1121-5.

[54]  Kadhim H, Tabarki B, De Prez C, Sebire G. Cytokine immunoreactivity in cortical and subcortical neurons in periventricular leukomalacia: are cytokines implicated in neuronal dysfunction in cerebral palsy? Acta Neuropathol. 2003 Mar;105(3):209-16.

[55]  Olah M, Amor S, Brouwer N, Vinet J, Eggen B, Biber K, *et al.* Identification of a microglia phenotype supportive of remyelination. Glia. 2012 Feb;60(2):306-21.

[56]  Carson MJB, Tina V; Puntambekar, Shweta S; Melchior, Benoit; Doose, Jonathan M; Ethell, Iryna M. Carson, Monica J; Bilousova, Tina V; Puntambekar, Shweta S; Melchior, Benoit; Doose, Jonathan M; Ethell, Iryna M. Neurotherapeutics : the journal of the American Society for Experimental NeuroTherapeutics. 2007;4(4):571-9.

[57]  Gehrmann J, Banati RB, Kreutzberg GW. Microglia in the Immune Surveillance of the Brain-Human Microglia Constitutively Express Hla-Dr Molecules. J Neuroimmunol. 1993 Nov-Dec;48(2):189-98.

[58]  Sasaki A, Nakazato Y. The Identity of Cells Expressing Mhc Class-Ii Antigens in Normal and Pathological Human Brain. Neuropathol Appl Neurobiol. 1992 Feb;18(1):13-26.

[59]  Ogura K, Ogawa M, Yoshida M. Effects of Aging on Microglia in the Normal Rat-Brain-Immunohistochemical Observations. Neuroreport. 1994 Jun 2;5(10):1224-6.

[60]  Ong WY, Leong SK, Garey LJ, Tan KK, Zhang HF. A Light and Electron-Microscopic Study of Hla-Dr Positive Cells in the Human Cerebral-Cortex and Subcortical White-Matter. Journal of Brain Research-Journal Fur Hirnforschung. 1995;36(4):553-63.

[61]  Alldinger S, Wunschmann A, Baumgartner W, Voss C, Kremmer E. Up-regulation of major histocompatibility complex class II antigen expression in the central nervous system of dogs with spontaneous canine distemper virus encephalitis. Acta Neuropathol (Berl). 1996 Sep;92(3):273-80.

[62]  Sheffield LG, Berman NEJ. Microglial expression of MHC class II increases in normal aging of nonhuman primates. Neurobiol Aging. 1998 Jan-Feb;19(1):47-55.

[63]  Styren SD, Civin WH, Rogers J. Molecular, cellular, and pathologic characterization of HLA-DR immunoreactivity in normal elderly and Alzheimer's disease brain. Exp Neurol. 1990 Oct;110(1):93-104.

[64]    Anderson AC, Anderson DE, Bregoli L, Hastings WD, Kassam N, Lei C, *et al.* Promotion of tissue inflammation by the immune receptor Tim-3 expressed on innate immune cells. Science. 2007 Nov 16;318(5853):1141-3.

[65]    Kullberg S, Aldskogius H, Ulfhake B. Microglial activation, emergence of ED1-expressing cells and clusterin upregulation in the aging rat CNS, with special reference to the spinal cord. Brain Res. 2001 Apr 27;899(1-2):169-86.

[66]    Wakita H, Tomimoto H, Akiguchi I, Kimura J. Glial Activation and White-Matter Changes in the Rat-Brain Induced by Chronic Cerebral Hypoperfusion-an Immunohistochemical Study. Acta Neuropathol (Berl). 1994 May;87(5):484-92.

[67]    Wakita H, Tomimoto H, Akiguchi I, Kimura J. Dose-dependent, protective effect of FK506 against white matter changes in the rat brain after chronic cerebral ischemia. Brain Res. 1998 May 4;792(1):105-13.

[68]    Wakita H, Tomimoto H, Akiguchi I, Kimura J. Protective effect of cyclosporin A on white matter changes in the rat brain after chronic cerebral hypoperfusion. Stroke. 1995 Aug;26(8):1415-22.

[69]    Ling EAN, Y K; Wu, C H; Kaur, C. Microglia: its development and role as a neuropathology sensor. Prog Brain Res. 2001;132:61-79.

[70]    Dheen ST, Kaur C, Ling EA. Microglial activation and its implications in the brain diseases. Curr Med Chem. 2007;14(11):1189-97.

[71]    Ling EA, Wong WC. The origin and nature of ramified and amoeboid microglia: a historical review and current concepts. Glia. 1993 Jan;7(1):9-18.

[72]    Imamoto K, Leblond CP. Radioautographic investigation of gliogenesis in the corpus callosum of young rats. II. Origin of microglial cells. J Comp Neurol. 1978 Jul 1;180(1):139-63.

[73]    Earle KL, Mitrofanis J. Identification of transient microglial cell colonies in the forebrain white matter of developing rats. J Comp Neurol. 1997 Oct 27;387(3):371-84.

[74]    Ling EA, Tan CK. Amoeboid microglial cells in the corpus callosum of neonatal rats. Arch Histol Jpn. 1974 Mar;36(4):265-80.

[75]    Rangarajan P, Jiang B, Baby N, Manivannan J, Jayapal M, Jia L, *et al.* Transcriptome analysis of amoeboid and ramified microglia isolated from the corpus callosum of rat brain. BMC Neurosci. 2012 Jun 14;13(1):64.

[76]    Rezaie P, Male D. Mesoglia & microglia--a historical review of the concept of mononuclear phagocytes within the central nervous system. J Hist Neurosci. 2002 Dec;11(4):325-74.

[77]    Levison SW, Rothstein RP, Brazel CY, Young GM, Albrecht PJ. Selective apoptosis within the rat subependymal zone: A plausible mechanism for determining which lineages develop from neural stem cells. Dev Neurosci. 2000 Jan-Apr;22(1-2):106-15.

[78]    Rakic S, Zecevic N. Programmed cell death in the developing human telencephalon. Eur J Neurosci. 2000 Aug;12(8):2721-34.

[79]    Haynes RL, Borenstein NS, Desilva TM, Folkerth RD, Liu LG, Volpe JJ, *et al.* Axonal development in the cerebral white matter of the human fetus and infant. J Comp Neurol. 2005 Apr 4;484(2):156-67.

[80]    Hamilton SP, Rome LH. Stimulation of *in Vitro* Myelin Synthesis by Microglia. Glia. 1994 Aug;11(4):326-35.

[81]    Witting A, Muller P, Herrmann A, Kettenmann H, Nolte C. Phagocytic clearance of apoptotic neurons by Microglia/Brain macrophages *in vitro*: involvement of lectin-,

integrin-, and phosphatidylserine-mediated recognition. J Neurochem. 2000 Sep;75(3):1060-70.

[82] Bodeutsch N, Thanos S. Migration of phagocytotic cells and development of the murine intraretinal microglial network: An *in vivo* study using fluorescent dyes. Glia. 2000 Oct;32(1):91-101.

[83] Innocenti GM, Clarke S. Multiple Sets of Visual Cortical-Neurons Projecting Transitorily through the Corpus-Callosum. Neurosci Lett. 1983;41(1-2):27-32.

[84] Ling EA, Kaur C, Wong WC. Light and electron microscopic demonstration of non-specific esterase in amoeboid microglial cells in the corpus callosum in postnatal rats: a cytochemical link to monocytes. J Anat. 1982 Sep;135(Pt 2):385-94.

[85] Ferrer I, Bernet E, Soriano E, del Rio T, Fonseca M. Naturally occurring cell death in the cerebral cortex of the rat and removal of dead cells by transitory phagocytes. Neuroscience. 1990;39(2):451-8.

[86] Ling EA. Some aspects of amoeboid microglia in the corpus callosum and neighbouring regions of neonatal rats. J Anat. 1976 Feb;121(Pt 1):29-45.

[87] Sondag CM, Dhawan G, Combs CK. Beta amyloid oligomers and fibrils stimulate differential activation of primary microglia. J Neuroinflammation. 2009;6:1.

[88] Voorend M, van der Ven AJ, Mulder M, Lodder J, Steinbusch HW, Bruggeman CA. Chlamydia pneumoniae infection enhances microglial activation in atherosclerotic mice. Neurobiol Aging. 2010 Oct;31(10):1766-73.

[89] Nakajima K, Kohsaka S, Tohyama Y, Kurihara T. Activation of microglia with lipopolysaccharide leads to the prolonged decrease of conventional protein kinase C activity. Brain Res Mol Brain Res. 2003 Jan 31;110(1):92-9.

[90] Abraham H, Losonczy A, Czeh G, Lazar G. Rapid activation of microglial cells by hypoxia, kainic acid, and potassium ions in slice preparations of the rat hippocampus. Brain Res. 2001 Jul 6;906(1-2):115-26.

[91] Gordon S. Alternative activation of macrophages. Nat Rev Immunol. 2003 Jan;3(1):23-35.

[92] Duffield JS. The inflammatory macrophage: a story of Jekyll and Hyde. Clin Sci. 2003 Jan;104(1):27-38.

[93] Goerdt S, Orfanos CE. Other functions, other genes: alternative activation of antigen-presenting cells. Immunity. 1999 Feb;10(2):137-42.

[94] Rathnasamy G, Ling EA, Kaur C. Microglia in the Developing Brain. Kaur CL, E A; Yong Loo Lin School of Medicine, National University of Singapore, editor. Newyork: Nova Science Publishers; 2012.

[95] Sanchez-Lopez AM, Cuadros MA, Calvente R, Tassi M, Marin-Teva JL, Navascues J. Activation of immature microglia in response to stab wound in embryonic quail retina. J Comp Neurol. 2005 Nov 7;492(1):20-33.

[96] Kaur C, Dheen ST, Ling EA. From blood to brain: amoeboid microglial cell, a nascent macrophage and its functions in developing brain. Acta Pharmacol Sin. 2007 Aug;28(8):1087-96.

[97] Kaur C, Too HF, Ling EA. Phagocytosis of Escherichia coli by amoeboid microglial cells in the developing brain. Acta Neuropathol. 2004 Mar;107(3):204-8.

[98] Kaur C, Ling EA, Wong WC. Transformation of amoeboid microglial cells into microglia in the corpus callosum of the postnatal rat brain. An electron microscopical study. Arch Histol Jpn. 1985 Feb;48(1):17-25.

[99]    Li YB, Kaur C, Ling EA. Neuronal degeneration and microglial reaction in the fetal and postnatal rat brain after transient maternal hypoxia. Neurosci Res. 1998 Oct;32(2):137-48.

[100]   Ling EA, Kaur C, Wong WC. Expression of major histocompatibility complex and leukocyte common antigens in amoeboid microglia in postnatal rats. J Anat. 1991 Aug;177:117-26.

[101]   Hayes GM, Woodroofe MN, Cuzner ML. Microglia are the major cell type expressing MHC class II in human white matter. J Neurol Sci. 1987 Aug;80(1):25-37.

[102]   Xu J, Ling EA. Upregulation and induction of surface antigens with special reference to MHC class II expression in microglia in postnatal rat brain following intravenous or intraperitoneal injections of lipopolysaccharide. J Anat. 1994 Apr;184 ( Pt 2):285-96.

[103]   Xu J, Ling EA. Upregulation and induction of major histocompatibility complex class I and II antigens on microglial cells in early postnatal rat brain following intraperitoneal injections of recombinant interferon-gamma. Neuroscience. 1994 Jun;60(4):959-67.

[104]   Kaur C, Sivakumar V, Dheen ST, Ling EA. Insulin-like growth factor I and II expression and modulation in amoeboid microglial cells by lipopolysaccharide and retinoic acid. Neuroscience. 2006;138(4):1233-44.

[105]   McRae A, Gilland E, Bona E, Hagberg H. Microglia activation after neonatal hypoxic-ischemia. Brain Res Dev Brain Res. 1995 Feb 16;84(2):245-52.

[106]   Morioka T, Kalehua AN, Streit WJ. Characterization of microglial reaction after middle cerebral artery occlusion in rat brain. J Comp Neurol. 1993 Jan 1;327(1):123-32.

[107]   Gehrmann J, Bonnekoh P, Miyazawa T, Hossmann KA, Kreutzberg GW. Immunocytochemical study of an early microglial activation in ischemia. J Cereb Blood Flow Metab. 1992 Mar;12(2):257-69.

[108]   Gehrmann J, Bonnekoh P, Miyazawa T, Oschlies U, Dux E, Hossmann KA, *et al.* The microglial reaction in the rat hippocampus following global ischemia: immuno-electron microscopy. Acta Neuropathol. 1992;84(6):588-95.

[109]   Morioka T, Kalehua AN, Streit WJ. Progressive expression of immunomolecules on microglial cells in rat dorsal hippocampus following transient forebrain ischemia. Acta Neuropathol. 1992;83(2):149-57.

[110]   Morioka T, Kalehua AN, Streit WJ. The microglial reaction in the rat dorsal hippocampus following transient forebrain ischemia. J Cereb Blood Flow Metab. 1991 Nov;11(6):966-73.

[111]   Tambuyzer BR, Ponsaerts P, Nouwen EJ. Microglia: gatekeepers of central nervous system immunology. J Leukoc Biol. 2009 Mar;85(3):352-70.

[112]   Hanisch UK, Prinz M, Angstwurm K, Hausler KG, Kann O, Kettenmann H, *et al.* The protein tyrosine kinase inhibitor AG126 prevents the massive microglial cytokine induction by pneumococcal cell walls. Eur J Immunol. 2001 Jul;31(7):2104-15.

[113]   Murugan M, Ling EA, Kaur C. Role of cytokines in hypoxic periventricular white matter damage in developing brain. Manjili MH, editor. Newyork: Nova Science Publishers; 2012.

[114]   Stanimirovic D, Zhang W, Howlett C, Lemieux P, Smith C. Inflammatory gene transcription in human astrocytes exposed to hypoxia: roles of the nuclear factor-kappaB and autocrine stimulation. J Neuroimmunol. 2001 Oct 1;119(2):365-76.

[115]   Hori O, Matsumoto M, Kuwabara K, Maeda Y, Ueda H, Ohtsuki T, *et al.* Exposure of astrocytes to hypoxia/reoxygenation enhances expression of glucose-regulated protein 78 facilitating astrocyte release of the neuroprotective cytokine interleukin 6. J Neurochem. 1996 Mar;66(3):973-9.

[116]   Cannella B, Raine CS. Multiple sclerosis: cytokine receptors on oligodendrocytes predict innate regulation. Ann Neurol. 2004 Jan;55(1):46-57.

[117] Sawada M, Itoh Y, Suzumura A, Marunouchi T. Expression of cytokine receptors in cultured neuronal and glial cells. Neurosci Lett. 1993 Oct 1;160(2):131-4.

[118] Omari KM, John GR, Sealfon SC, Raine CS. CXC chemokine receptors on human oligodendrocytes: implications for multiple sclerosis. Brain. 2005 May;128(Pt 5):1003-15.

[119] Jones SJ, Ledgerwood EC, Prins JB, Galbraith J, Johnson DR, Pober JS, *et al*. TNF recruits TRADD to the plasma membrane but not the trans-Golgi network, the principal subcellular location of TNF-R1. Journal of immunology (Baltimore, Md : 1950). 1999;162(2):1042-8.

[120] Ichijo H, Nishida E, Irie K, tenDijke P, Saitoh M, Moriguchi T, *et al*. Induction of apoptosis by ASK1, a mammalian MAPKKK that activates SAPK/JNK and p38 signaling pathways. Science. 1997 Jan 3;275(5296):90-4.

[121] Vela JM, Molina-Holgado E, Arevalo-Martin A, Almazan G, Guaza C. Interleukin-1 regulates proliferation and differentiation of oligodendrocyte progenitor cells. Mol Cell Neurosci. 2002 Jul;20(3):489-502.

[122] Akassoglou K, Bauer J, Kassiotis G, Pasparakis M, Lassmann H, Kollias G, *et al*. Oligodendrocyte apoptosis and primary demyelination induced by local TNF/p55TNF receptor signaling in the central nervous system of transgenic mice: models for multiple sclerosis with primary oligodendrogliopathy. Am J Pathol. 1998 Sep;153(3):801-13.

[123] Lock C, Oksenberg J, Steinman L. The role of TNFalpha and lymphotoxin in demyelinating disease. Ann Rheum Dis. 1999 Nov;58 Suppl 1:I121-8.

[124] Chatterjee S. Neutral sphingomyelinase action stimulates signal transduction of tumor necrosis factor-alpha in the synthesis of cholesteryl esters in human fibroblasts. J Biol Chem. 1994 Jan 14;269(2):879-82.

[125] Ledeen RW, Chakraborty G. Cytokines, signal transduction, and inflammatory demyelination: review and hypothesis. Neurochem Res. 1998 Mar;23(3):277-89.

[126] Kremlev SG, Roberts RL, Palmer C. Differential expression of chemokines and chemokine receptors during microglial activation and inhibition. J Neuroimmunol. 2004 Apr;149(1-2):1-9.

[127] Wang X, Li C, Chen Y, Hao Y, Zhou W, Chen C, *et al*. Hypoxia enhances CXCR4 expression favoring microglia migration *via* HIF-1alpha activation. Biochem Biophys Res Commun. 2008 Jun 27;371(2):283-8.

[128] Sunnemark D, Eltayeb S, Wallstrom E, Appelsved L, Malmberg A, Lassmann H, *et al*. Differential expression of the chemokine receptors CX3CR1 and CCR1 by microglia and macrophages in myelin-oligodendrocyte-glycoprotein-induced experimental autoimmune encephalomyelitis. Brain Pathol. 2003 Oct;13(4):617-29.

[129] Flynn G, Maru S, Loughlin J, Romero IA, Male D. Regulation of chemokine receptor expression in human microglia and astrocytes. J Neuroimmunol. 2003 Mar;136(1-2):84-93.

[130] Dorf ME, Berman MA, Tanabe S, Heesen M, Luo Y. Astrocytes express functional chemokine receptors. J Neuroimmunol. 2000 Nov 1;111(1-2):109-21.

[131] Aravalli RN, Hu S, Rowen TN, Palmquist JM, Lokensgard JR. Cutting edge: TLR2-mediated pro-inflammatory cytokine and chemokine production by microglial cells in response to herpes simplex virus. J Immunol. 2005 Oct 1;175(7):4189-93.

[132] Tomita MH, Brita J; Santoro, Christopher P; Santoro, Thomas J. Astrocyte production of the chemokine macrophage inflammatory protein-2 is inhibited by the spice principle curcumin at the level of gene transcription. J Neuroinflammation. 2005;2(1):8.

[133] Takanohashi A, Yabe T, Schwartz JP. Pigment epithelium-derived factor induces the production of chemokines by rat microglia. Glia. 2005 Sep;51(4):266-78.

[134] Rezaie P, Trillo-Pazos G, Everall IP, Male DK. Expression of beta-chemokines and chemokine receptors in human fetal astrocyte and microglial co-cultures: Potential role of chemokines in the developing CNS. Glia. 2002 Jan;37(1):64-75.

[135] Salmaggi A, Gelati M, Dufour A, Corsini E, Pagano S, Baccalini R, *et al.* Expression and modulation of IFN-gamma-Inducible chemokines (IP-10, Mig, and I-TAC) in human brain endothelium and astrocytes: Possible relevance for the immune invasion of the central nervous system and the pathogenesis of multiple sclerosis. J Interferon Cytokine Res. 2002 Jun;22(6):631-40.

[136] Cole KE, Strick CA, Paradis TJ, Ogborne KT, Loetscher M, Gladue RP, *et al.* Interferon-inducible T cell alpha chemoattractant (I-TAC): A novel non-ELR CXC chemokine with potent activity on activated T cells through selective high affinity binding to CXCR3. J Exp Med. 1998 Jun 15;187(12):2009-21.

[137] Cross AK, Woodroofe MN. Chemokines induce migration and changes in actin polymerization in adult rat brain microglia and a human fetal microglial cell line *in vitro*. J Neurosci Res. 1999 Jan 1;55(1):17-23.

[138] Ambrosini E, Aloisi F. Chemokines and glial cells: a complex network in the central nervous system. Neurochem Res. 2004 May;29(5):1017-38.

[139] Silverstein FS, Naik B, Simpson J. Hypoxia-ischemia stimulates hippocampal glutamate efflux in perinatal rat brain: an *in vivo* microdialysis study. Pediatr Res. 1991 Dec;30(6):587-90.

[140] Benveniste H, Drejer J, Schousboe A, Diemer NH. Elevation of the extracellular concentrations of glutamate and aspartate in rat hippocampus during transient cerebral ischemia monitored by intracerebral microdialysis. J Neurochem. 1984 Nov;43(5):1369-74.

[141] Hagberg H. Hypoxic-Ischemic Damage in the Neonatal Brain-Excitatory Amino-Acids. Dev Pharmacol Ther. 1992;18(3-4):139-44.

[142] Patrizio M, Levi G. Glutamate Production by Cultured Microglia-Differences between Rat and Mouse, Enhancement by Lipopolysaccharide and Lack Effect of Hiv Coat Protein Gp120 and Depolarizing Agents. Neurosci Lett. 1994 Sep 12;178(2):184-8.

[143] Sivakumar V, Ling EA, Liu J, Kaur C. Role of Glutamate and Its Receptors and Insulin-like Growth Factors in Hypoxia Induced Periventricular White Matter Injury. Glia. 2010 Apr 1;58(5):507-23.

[144] Domercq M, Sanchez-Gomez MV, Sherwin C, Etxebarria E, Fern R, Matute C. System xc-and glutamate transporter inhibition mediates microglial toxicity to oligodendrocytes. J Immunol. 2007 May 15;178(10):6549-56.

[145] Barger SW, Goodwin ME, Porter MM, Beggs ML. Glutamate release from activated microglia requires the oxidative burst and lipid peroxidation. J Neurochem. 2007 Jun;101(5):1205-13.

[146] Raymond M, Li PJ, Mangin JM, Huntsman M, Gallo V. Chronic Perinatal Hypoxia Reduces Glutamate-Aspartate Transporter Function in Astrocytes through the Janus Kinase/Signal Transducer and Activator of Transcription Pathway. J Neurosci. 2011 Dec 7;31(49):17864-71.

[147] Kinney HC, Back SA. Human oligodendroglial development: relationship to periventricular leukomalacia. Semin Pediatr Neurol. 1998 Sep;5(3):180-9.

[148] Yoshioka A, Bacskai B, Pleasure D. Pathophysiology of oligodendroglial excitotoxicity. J Neurosci Res. 1996 Nov 15;46(4):427-37.

[149]  Oka A, Belliveau MJ, Rosenberg PA, Volpe JJ. Vulnerability of oligodendroglia to glutamate: pharmacology, mechanisms, and prevention. J Neurosci. 1993 Apr;13(4):1441-53.

[150]  Yoshioka A, Hardy M, Younkin DP, Grinspan JB, Stern JL, Pleasure D. Alpha-amino-3-hydroxy-5-methyl-4-isoxazolepropionate (AMPA) receptors mediate excitotoxicity in the oligodendroglial lineage. J Neurochem. 1995 Jun;64(6):2442-8.

[151]  Matute C, Sanchez-Gomez MV, Martinez-Millan L, Miledi R. Glutamate receptor-mediated toxicity in optic nerve oligodendrocytes. Proc Natl Acad Sci U S A. 1997 Aug 5;94(16):8830-5.

[152]  McDonald JW, Althomsons SP, Hyrc KL, Choi DW, Goldberg MP. Oligodendrocytes from forebrain are highly vulnerable to AMPA/kainate receptor-mediated excitotoxicity. Nat Med. 1998 Mar;4(3):291-7.

[153]  Pitt D, Werner P, Raine CS. Glutamate excitotoxicity in a model of multiple sclerosis. Nat Med. 2000 Jan;6(1):67-70.

[154]  Grunder T, Kohler K, Guenther E. Alterations in NMDA receptor expression during retinal degeneration in the RCS rat. Vis Neurosci. 2001 Sep-Oct;18(5):781-7.

[155]  Volpe JJ. Neurobiology of periventricular leukomalacia in the premature infant. Pediatr Res. 2001 Nov;50(5):553-62.

[156]  Noda M, Nakanishi H, Nabekura J, Akaike N. AMPA-kainate subtypes of glutamate receptor in rat cerebral microglia. J Neurosci. 2000 Jan 1;20(1):251-8.

[157]  Hagino Y, Kariura Y, Manago Y, Amano T, Wang B, Sekiguchi M, *et al*. Heterogeneity and potentiation of AMPA type of glutamate receptors in rat cultured microglia. Glia. 2004 Jul;47(1):68-77.

[158]  Selmaj KW, Raine CS. Tumor necrosis factor mediates myelin and oligodendrocyte damage *in vitro*. Ann Neurol. 1988 Apr;23(4):339-46.

[159]  Gottlieb M, Matute C. Expression of ionotropic glutamate receptor subunits in glial cells of the hippocampal CA1 area following transient forebrain ischemia. J Cereb Blood Flow Metab. 1997 Mar;17(3):290-300.

[160]  Lipton SA. Pathologically activated therapeutics for neuroprotection (Vol 8, pg 803, 2007). Nature Reviews Neuroscience. 2007 Nov;8(11).

[161]  Pocock JM, Kettenmann H. Neurotransmitter receptors on microglia. Trends Neurosci. 2007 Oct;30(10):527-35.

[162]  Taylor DL, Jones F, Kubota ESFCS, Pocock JM. Stimulation of microglial metabotropic glutamate receptor mGlu2 triggers tumor necrosis factor alpha-induced neurotoxicity in concert with microglial-derived fas ligand. J Neurosci. 2005 Mar 16;25(11):2952-64.

[163]  Taylor DL, Diemel LT, Pocock JM. Activation of microglial group III metabotropic glutamate receptors protects neurons against microglial neurotoxicity. J Neurosci. 2003 Mar 15;23(6):2150-60.

[164]  Biber K, Laurie DJ, Berthele A, Sommer B, Tolle TR, Gebicke-Harter PJ, *et al*. Expression and signaling of group I metabotropic glutamate receptors in astrocytes and microglia. J Neurochem. 1999 Apr;72(4):1671-80.

[165]  Taylor DL, Diemel LT, Cuzner ML, Pocock JM. Activation of group II metabotropic glutamate receptors underlies microglial reactivity and neurotoxicity following stimulation with chromogranin A, a peptide up-regulated in Alzheimer's disease. J Neurochem. 2002 Sep;82(5):1179-91.

[166]   Kingham PJP, J M. Microglial apoptosis induced by chromogranin A is mediated by mitochondrial depolarisation and the permeability transition but not by cytochrome c release. J Neurochem. 2000;74(4):1452-62.

[167]   Ferriero DM. Oxidant mechanisms in neonatal hypoxia-ischemia. Dev Neurosci. 2001;23(3):198-202.

[168]   Connor JR, Menzies SL. Relationship of iron to oligodendrocytes and myelination. Glia. 1996 Jun;17(2):83-93.

[169]   Beard JL, Connor JR. Iron status and neural functioning. Annu Rev Nutr. 2003;23:41-58.

[170]   Cheepsunthorn P, Palmer C, Connor JR. Cellular distribution of ferritin subunits in postnatal rat brain. J Comp Neurol. 1998 Oct 12;400(1):73-86.

[171]   Connor JR, Menzies SL. Cellular management of iron in the brain. J Neurol Sci. 1995 Dec;134:33-44.

[172]   Kaur C, Ling EA. Increased expression of transferrin receptors and iron in amoeboid microglial cells in postnatal rats following an exposure to hypoxia. Neurosci Lett. 1999 Mar 12;262(3):183-6.

[173]   Mehlhase J, Sandig G, Pantopoulos K, Grune T. Oxidation-induced ferritin turnover in microglial cells: role of proteasome. Free Radic Biol Med. 2005 Jan 15;38(2):276-85.

[174]   Gorter JA, Mesquita ARM, van Vliet EA, da Silva FHL, Aronica E. Increased expression of ferritin, an iron-storage protein, in specific regions of the parahippocampal cortex of epileptic rats. Epilepsia. 2005 Sep;46(9):1371-9.

[175]   Berg D, Gerlach M, Youdim MBH, Double KL, Zecca L, Riederer P, *et al*. Brain iron pathways and their relevance to Parkinson's disease. J Neurochem. 2001 Oct;79(2):225-36.

[176]   Adcock LM, Yamashita Y, GoddardFinegold J, Smith CV. Cerebral hypoxia-ischemia increases microsomal iron in newborn piglets. Metab Brain Dis. 1996 Dec;11(4):359-67.

[177]   Thompson KJ, Shoham S, Connor JR. Iron and neurodegenerative disorders. Brain Res Bull. 2001 May 15;55(2):155-64.

[178]   Craelius W, Migdal MW, Luessenhop CP, Sugar A, Mihalakis I. Iron Deposits Surrounding Multiple-Sclerosis Plaques. Arch Pathol Lab Med. 1982;106(8):397-9.

[179]   Connor JR, Menzies SL. Altered Cellular-Distribution of Iron in the Central-Nervous-System of Myelin Deficient Rats. Neuroscience. 1990;34(1):265-71.

[180]   Zhang X, Surguladze N, Slagle-Webb B, Connor JR. Cellular iron status influences the functional relationship between microglia and Oligodendrocytes. Glia. 2006 Dec;54(8):795-804.

[181]   Saleppico S, Mazzolla R, Boelaert JR, Puliti M, Barluzzi R, Bistoni F, *et al*. Iron regulates microglial cell-mediated secretory and effector functions. Cellular Immunology. 1996 Jun 15;170(2):251-9.

[182]   Kadhim H, Khalifa M, Deltenre P, Casimir G, Sebire G. Molecular mechanisms of cell death in periventricular leukomalacia. Neurology. 2006 Jul 25;67(2):293-9.

[183]   Mishra OP, Zanelli S, Ohnishi ST, Delivoria-Papadopoulos M. Hypoxia-induced generation of nitric oxide free radicals in cerebral cortex of newborn guinea pigs. Neurochem Res. 2000 Dec;25(12):1559-65.

[184]   Mishra OP, Mishra R, Ashraf QM, Delivoria-Papadopoulos M. Nitric oxide-mediated mechanism of neuronal nitric oxide synthase and inducible nitric oxide synthase expression during hypoxia in the cerebral cortex of newborn piglets. Neuroscience. 2006;140(3):857-63.

[185]   van den Tweel ERW, Nijboer C, Kavelaars A, Heijnen CJ, Groenendaal F, van Bel F. Expression of nitric oxide synthase isoforms and nitrotyrosine formation after hypoxia-ischemia in the neonatal rat brain. J Neuroimmunol. 2005 Oct;167(1-2):64-71.

[186] Dong Y, Yu Z, Sun Y, Zhou H, Stites J, Newell K, *et al.* Chronic fetal hypoxia produces selective brain injury associated with altered nitric oxide synthases. Am J Obstet Gynecol. 2011 Mar;204(3):254 e16-28.

[187] Spandou E, Karkavelas G, Soubasi V, Avgovstides-Savvopoulou P, Loizidis T, Guiba-Tziampiri O. Effect of ketamine on hypoxic-ischemic brain damage in newborn rats. Brain Res. 1999 Feb 20;819(1-2):1-7.

[188] Lu D-YL, Houng-Chi; Tang, Chih-Hsin; Fu, Wen-Mei. Hypoxia-induced iNOS expression in microglia is regulated by the PI3-kinase/Akt/mTOR signaling pathway and activation of hypoxia inducible factor-1alpha. Biochem Pharmacol. 2006;72(8):992-1000.

[189] Baud O, Li JR, Zhang YM, Neve RL, Volpe JJ, Rosenberg PA. Nitric oxide-induced cell death in developing oligodendrocytes is associated with mitochondrial dysfunction and apoptosis-inducing factor translocation. Eur J Neurosci. 2004 Oct;20(7):1713-26.

[190] Peeters-Scholte C, Koster J, Veldhuis W, van den Tweel E, Zhu CL, Kops N, *et al.* Neuroprotection by selective nitric oxide synthase inhibition at 24 hours after perinatal hypoxia-ischemia. Stroke. 2002 Sep;33(9):2304-10.

[191] Kaur C, Sivakumar V, Yip GW, Ling EA. Expression of syndecan-2 in the amoeboid microglial cells and its involvement in inflammation in the hypoxic developing brain. Glia. 2009 Feb;57(3):336-49.

[192] Welin AK, Sandberg M, Lindblom A, Arvidsson P, Nilsson UA, Kjellmer I, *et al.* White matter injury following prolonged free radical formation in the 0.65 gestation fetal sheep brain. Pediatr Res. 2005 Jul;58(1):100-5.

[193] Sankarapandi S, Zweier JL, Mukherjee G, Quinn MT, Huso DL. Measurement and characterization of superoxide generation in microglial cells: evidence for an NADPH oxidase-dependent pathway. Arch Biochem Biophys. 1998 May 15;353(2):312-21.

[194] Zekry D, Epperson TK, Krause KH. A role for NOX NADPH oxidases in Alzheimer's disease and other types of dementia? IUBMB Life. 2003 Jun;55(6):307-13.

[195] Lavigne MC, Malech HL, Holland SM, Leto TL. Genetic requirement of p47phox for superoxide production by murine microglia. FASEB J. 2001 Feb;15(2):285-7.

[196] Green SP, Cairns B, Rae J, Errett-Baroncini C, Hongo JA, Erickson RW, *et al.* Induction of gp91-phox, a component of the phagocyte NADPH oxidase, in microglial cells during central nervous system inflammation. J Cereb Blood Flow Metab. 2001 Apr;21(4):374-84.

[197] Infanger DW, Sharma RV, Davisson RL. NADPH oxidases of the brain: distribution, regulation, and function. Antioxid Redox Signal. 2006 Sep-Oct;8(9-10):1583-96.

[198] Qin L, Liu Y, Wang T, Wei SJ, Block ML, Wilson B, *et al.* NADPH oxidase mediates lipopolysaccharide-induced neurotoxicity and pro-inflammatory gene expression in activated microglia. J Biol Chem. 2004 Jan 9;279(2):1415-21.

[199] Olivier P, Fontaine RH, Loron G, Van Steenwinckel J, Biran V, Massonneau V, *et al.* Melatonin Promotes Oligodendroglial Maturation of Injured White Matter in Neonatal Rats. PLoS ONE. 2009 Sep 22;4(9).

[200] Meda L, Baron P, Scarlato G. Glial activation in Alzheimer's disease: the role of Abeta and its associated proteins. Neurobiol Aging. 2001 Nov-Dec;22(6):885-93.

[201] Czlonkowska A, Kurkowska-Jastrzebska I, Czlonkowski A, Peter D, Stefano GB. Immune processes in the pathogenesis of Parkinson's disease-a potential role for microglia and nitric oxide. Med Sci Monit. 2002 Aug;8(8):RA165-77.

[202] Monji A, Kato T, Kanba S. Cytokines and schizophrenia: Microglia hypothesis of schizophrenia. Psychiatry Clin Neurosci. 2009 Jun;63(3):257-65.

[203] Hu S, Chao CC, Khanna KV, Gekker G, Peterson PK, Molitor TW. Cytokine and free radical production by porcine microglia. Clin Immunol Immunopathol. 1996 Jan;78(1):93-6.

[204] Iiyama M, Kakihana K, Kurosu T, Miura O. Reactive oxygen species generated by hematopoietic cytokines play roles in activation of receptor-mediated signaling and in cell cycle progression. Cell Signal. 2006 Feb;18(2):174-82.

[205] Sattler M, Winkler T, Verma S, Byrne CH, Shrikhande G, Salgia R, *et al*. Hematopoietic growth factors signal through the formation of reactive oxygen species. Blood. 1999 May 1;93(9):2928-35.

[206] Ishizuka YA, Hiroshi; Nakane, Hideyuki; Kannan, Hiroshi; Ishida, Yasushi. Different response between production of free radicals induced by central and peripheral administration of interleukin-1beta in conscious rats. Neurosci Res. 2008;60(1):10-4.

[207] Lu R, Wang P, Wartofsky L, Sutton BD, Zweier JL, Bahn RS, *et al*. Oxygen free radicals in interleukin-1beta-induced glycosaminoglycan production by retro-ocular fibroblasts from normal subjects and Graves' ophthalmopathy patients. Thyroid. 1999 Mar;9(3):297-303.

[208] Parathath SR, Parathath S, Tsirka SE. Nitric oxide mediates neurodegeneration and breakdown of the blood-brain barrier in tPA-dependent excitotoxic injury in mice. J Cell Sci. 2006 Jan 15;119(Pt 2):339-49.

[209] Schreibelt G, Kooij G, Reijerkerk A, van Doorn R, Gringhuis SI, van der Pol S, *et al*. Reactive oxygen species alter brain endothelial tight junction dynamics *via* RhoA, PI3 kinase, and PKB signaling. FASEB J. 2007 Nov;21(13):3666-76.

[210] Phares TW, Fabis MJ, Brimer CM, Kean RB, Hooper DC. A peroxynitrite-dependent pathway is responsible for blood-brain barrier permeability changes during a central nervous system inflammatory response: TNF-alpha is neither necessary nor sufficient. J Immunol. 2007 Jun 1;178(11):7334-43.

[211] Haynes RL, Baud O, Li J, Kinney HC, Volpe JJ, Folkerth DR. Oxidative and nitrative injury in periventricular leukomalacia: a review. Brain Pathol. 2005 Jul;15(3):225-33.

[212] Buonocore G, Perrone S, Bracci R. Free radicals and brain damage in the newborn. Biol Neonate. 2001;79(3-4):180-6.

[213] Wallin C, Puka-Sundvall M, Hagberg H, Weber SG, Sandberg M. Alterations in glutathione and amino acid concentrations after hypoxia-ischemia in the immature rat brain. Brain Res Dev Brain Res. 2000 Dec 29;125(1-2):51-60.

[214] Colton C, Wilt S, Gilbert D, Chernyshev O, Snell J, Dubois-Dalcq M. Species differences in the generation of reactive oxygen species by microglia. Mol Chem Neuropathol. 1996 May-Aug;28(1-3):15-20.

[215] Zhang N, Komine-Kobayashi M, Tanaka R, Liu M, Mizuno Y, Urabe T. Edaravone reduces early accumulation of oxidative products and sequential inflammatory responses after transient focal ischemia in mice brain. Stroke. 2005 Oct;36(10):2220-5.

[216] Nakanishi H. Neuronal and microglial cathepsins in aging and age-related diseases. Ageing Res Rev. 2003 Oct;2(4):367-81.

[217] del Zoppo GJ, Milner R, Mabuchi T, Hung S, Wang X, Berg GI, *et al*. Microglial activation and matrix protease generation during focal cerebral ischemia. Stroke. 2007 Feb;38(2 Suppl):646-51.

[218] Chen J, Cui X, Zacharek A, Cui Y, Roberts C, Chopp M. White matter damage and the effect of matrix metalloproteinases in type 2 diabetic mice after stroke. Stroke. 2011 Feb;42(2):445-52.

[219]   Maeda A, Sobel RA. Matrix metalloproteinases in the normal human central nervous system, microglial nodules, and multiple sclerosis lesions. J Neuropath Exp Neur. 1996 Mar;55(3):300-9.

[220]   Ito S, Kimura K, Haneda M, Ishida Y, Sawada M, Isobe K. Induction of matrix metalloproteinases (MMP3, MMP12 and MMP13) expression in the microglia by amyloid-beta stimulation *via* the PI3K/Akt pathway. Exp Gerontol. 2007 Jun;42(6):532-7.

[221]   Yamada T, Yoshiyama Y, Sato H, Seiki M, Shinagawa A, Takahashi M. White matter microglia produce membrane-type matrix metalloprotease, an activator of gelatinase A, in human brain tissues. Acta Neuropathol. 1995;90(5):421-4.

[222]   Svedin P, Hagberg H, Savman K, Zhu CL, Mallard C. Matrix metalloproteinase-9 gene knock-out protects the immature brain after cerebral hypoxia-ischemia. J Neurosci. 2007 Feb 14;27(7):1511-8.

[223]   Nakajima K, Shimojo M, Hamanoue M, Ishiura S, Sugita H, Kohsaka S. Identification of Elastase as a Secretory Protease from Cultured Rat Microglia. J Neurochem. 1992 Apr;58(4):1401-8.

[224]   Nakajima K, Nagata K, Hamanoue M, Takemoto N, Kohsaka S. Microglia-Derived Elastase Produces a Low-Molecular-Weight Plasminogen That Enhances Neurite Outgrowth in Rat Neocortical Explant Cultures. J Neurochem. 1993 Dec;61(6):2155-63.

[225]   Coan EJ, Saywood W, Collingridge GL. MK-801 blocks NMDA receptor-mediated synaptic transmission and long term potentiation in rat hippocampal slices. Neurosci Lett. 1987 Sep 11;80(1):111-4.

[226]   Thomas DM, Kuhn DM. MK-801 and dextromethorphan block microglial activation and protect against methamphetamine-induced neurotoxicity. Brain Res. 2005 Jul 19;1050(1-2):190-8.

[227]   Streit WJ, Morioka T, Kalehua AN. Mk-801 Prevents Microglial Reaction in Rat Hippocampus after Forebrain Ischemia. Neuroreport. 1992 Feb;3(2):146-8.

[228]   Bonnier C, Mesples B, Carpentier S, Henin D, Gressens P. Delayed white matter injury in a murine model of shaken baby syndrome. Brain Pathol. 2002 Jul;12(3):320-8.

[229]   Hirayama M, Kuriyama M. MK-801 is cytotoxic to microglia *in vitro* and its cytotoxicity is attenuated by glutamate, other excitotoxic agents and atropine. Possible presence of glutamate receptor and muscarinic receptor on microglia. Brain Res. 2001 Apr 6;897(1-2):204-6.

[230]   Han RZ, Hu JJ, Weng YC, Li DF, Huang Y. NMDA receptor antagonist MK-801 reduces neuronal damage and preserves learning and memory in a rat model of traumatic brain injury. Neuroscience Bulletin. 2009 Dec;25(6):367-75.

[231]   Park GR, Manara AR, Mendel L, Bateman PE. Ketamine Infusion-Its Use as a Sedative, Inotrope and Bronchodilator in a Critically Ill Patient. Anaesthesia. 1987 Sep;42(9):980-3.

[232]   Kawasaki T, Ogata M, Kawasaki C, Ogata J, Inoue Y, Shigematsu A. Ketamine suppresses pro-inflammatory cytokine production in human whole blood *in vitro*. Anesth Analg. 1999 Sep;89(3):665-9.

[233]   Shimaoka M, Iida T, Ohara A, Taenaka N, Mashimo T, Honda T, *et al*. Ketamine inhibits nitric oxide production in mouse-activated macrophage-like cells. Br J Anaesth. 1996 Aug;77(2):238-42.

[234]   Koga K, Ogata M, Takenaka I, Matsumoto T, Shigematsu A. Ketamine Suppresses Tumor-Necrosis-Factor-Alpha Activity and Mortality in Carrageenan-Sensitized Endotoxin-Shock Model. Circ Shock. 1994 Nov;44(3):160-8.

[235]  Shibakawa YS, Sasaki Y, Goshima Y, Echigo N, Kamiya Y, Kurahashi K, *et al.* Effects of ketamine and propofol on inflammatory responses of primary glial cell cultures stimulated with lipopolysaccharide. Br J Anaesth. 2005 Dec;95(6):803-10.

[236]  Chang Y, Lee JJ, Hsieh CY, Hsiao G, Chou DS, Sheu JR. Inhibitory Effects of Ketamine on Lipopolysaccharide-Induced Microglial Activation. Mediators Inflamm. 2009.

[237]  Howes MC. Ketamine for paediatric sedation/analgesia in the emergency department. Emerg Med J. 2004 May;21(3):275-80.

[238]  De Kock M. Ketamine for paediatric anaesthesia: useful, fool or... ? Annales Francaises D Anesthesie Et De Reanimation. 2007 Jun;26(6):524-8.

[239]  Bergman SA. Ketamine: review of its pharmacology and its use in pediatric anesthesia. Anesth Prog. 1999;46(1):10-20.

[240]  Danysz W, Lanthorn TH. Uncompetitive NMDA receptor antagonists--is low affinity better? Amino Acids. 2000;19(1):131-2.

[241]  Chen HS, Lipton SA. The chemical biology of clinically tolerated NMDA receptor antagonists. J Neurochem. 2006 Jun;97(6):1611-26.

[242]  Rogawski MA. Low affinity channel blocking (uncompetitive) NMDA receptor antagonists as therapeutic agents-toward an understanding of their favorable tolerability. Amino Acids. 2000;19(1):133-49.

[243]  Kumar S. Memantine: pharmacological properties and clinical uses. Neurol India. 2004 Sep;52(3):307-9.

[244]  Rosi S, Vazdarjanova A, Ramirez-Amaya V, Worley PF, Barnes CA, Wenk GL. Memantine protects against LPS-induced neuroinflammation, restores behaviorally-induced gene expression and spatial learning in the rat. Neuroscience. 2006 Nov 3;142(4):1303-15.

[245]  Takeda K, Muramatsu M, Chikuma T, Kato T. Effect of memantine on the levels of neuropeptides and microglial cells in the brain regions of rats with neuropathic pain. J Mol Neurosci. 2009 Nov;39(3):380-90.

[246]  Wu HM, Tzeng NS, Qian L, Wei SJ, Hu X, Chen SH, *et al.* Novel neuroprotective mechanisms of memantine: increase in neurotrophic factor release from astroglia and anti-inflammation by preventing microglial activation. Neuropsychopharmacology. 2009 Sep;34(10):2344-57.

[247]  Manning SM, Talos DM, Zhou C, Selip DB, Park HK, Park CJ, *et al.* NMDA receptor blockade with memantine attenuates white matter injury in a rat model of periventricular leukomalacia. J Neurosci. 2008 Jun 25;28(26):6670-8.

[248]  Li WJ, Chen HJ, Qian LH, He YF, Chen GY. [Effects of glial cell line-derived neurotrophic factor and memantine on long-term prognosis in neonatal rats with periventricular leukomalacia]. Zhongguo Dang Dai Er Ke Za Zhi. 2011 Sep;13(9):743-6.

[249]  Reisberg B, Doody R, Stoffler A, Schmitt F, Ferris S, Mobius HJ. Memantine in moderate-to-severe Alzheimer's disease. N Engl J Med. 2003 Apr 3;348(14):1333-41.

[250]  Howard R, McShane R, Lindesay J, Ritchie C, Baldwin A, Barber R, *et al.* Donepezil and memantine for moderate-to-severe Alzheimer's disease. N Engl J Med. 2012 Mar 8;366(10):893-903.

[251]  Balchen T, Diemer NH. The Ampa Antagonist, Nbqx, Protects against Ischemia-Induced Loss of Cerebellar Purkinje-Cells. Neuroreport. 1992 Sep;3(9):785-8.

[252]  Alford S, Grillner S. CNQX and DNQX block non-NMDA synaptic transmission but not NMDA-evoked locomotion in lamprey spinal cord. Brain Res. 1990;506(2):297-302.

[253] Suma T, Koshinaga M, Fukushima M, Kano T, Katayama Y. Effects of *in situ* administration of excitatory amino acid antagonists on rapid microglial and astroglial reactions in rat hippocampus following traumatic brain injury. Neurol Res. 2008 May;30(4):420-9.

[254] Ben Achour S, Pascual O. Glia: the many ways to modulate synaptic plasticity. Neurochem Int. 2010 Nov;57(4):440-5.

[255] Follett PL, Rosenberg PA, Volpe JJ, Jensen FE. NBQX attenuates excitotoxic injury in developing white matter. J Neurosci. 2000 Dec 15;20(24):9235-41.

[256] Takahashi JL, Giuliani F, Power C, Imai Y, Yong VW. Interleukin-1beta promotes oligodendrocyte death through glutamate excitotoxicity. Ann Neurol. 2003;53(5):588-95.

[257] Angehagen M, Ronnback L, Hansson E, Ben-Menachem E. Topiramate reduces AMPA-induced Ca(2+) transients and inhibits GluR1 subunit phosphorylation in astrocytes from primary cultures. J Neurochem. 2005 Aug;94(4):1124-30.

[258] Sfaello I, Baud O, Arzimanoglou A, Gressens P. Topiramate prevents excitotoxic damage in the newborn rodent brain. Neurobiol Dis. 2005 Dec;20(3):837-48.

[259] Shank RP, Gardocki JF, Vaught JL, Davis CB, Schupsky JJ, Raffa RB, *et al.* Topiramate: preclinical evaluation of structurally novel anticonvulsant. Epilepsia. 1994 Mar-Apr;35(2):450-60.

[260] Petrat F, de Groot H, Sustmann R, Rauen U. The chelatable iron pool in living cells: a methodically defined quantity. Biol Chem. 2002 Mar-Apr;383(3-4):489-502.

[261] Halliwell B, Gutteridge JM. Role of free radicals and catalytic metal ions in human disease: an overview. Methods Enzymol. 1990;186:1-85.

[262] Kaur D, Yantiri F, Rajagopalan S, Kumar J, Mo JQ, Boonplueang R, *et al.* Genetic or pharmacological iron chelation prevents MPTP-induced neurotoxicity *in vivo*: a novel therapy for Parkinson's disease. Neuron. 2003 Mar 27;37(6):899-909.

[263] Alam ZID, S E; Lees, A J; Marsden, D C; Jenner, P; Halliwell, B. A generalised increase in protein carbonyls in the brain in Parkinson's but not incidental Lewy body disease. J Neurochem. 1997;69(3):1326-9.

[264] Dexter DT, Statton SA, Whitmore C, Freinbichler W, Weinberger P, Tipton KF, *et al.* Clinically available iron chelators induce neuroprotection in the 6-OHDA model of Parkinson's disease after peripheral administration. J Neural Transm. 2011 Feb;118(2):223-31.

[265] Cherny RA, Atwood CS, Xilinas ME, Gray DN, Jones WD, McLean CA, *et al.* Treatment with a copper-zinc chelator markedly and rapidly inhibits beta-amyloid accumulation in Alzheimer's disease transgenic mice. Neuron. 2001 Jun;30(3):665-76.

[266] Peeters-Scholte C, Braun K, Koster J, Kops N, Blomgren K, Buonocore G, *et al.* Effects of allopurinol and deferoxamine on reperfusion injury of the brain in newborn piglets after neonatal hypoxia-ischemia. Pediatr Res. 2003 Oct;54(4):516-22.

[267] Palmer C, Roberts RL, Bero C. Deferoxamine posttreatment reduces ischemic brain injury in neonatal rats. Stroke. 1994 May;25(5):1039-45.

[268] Hurn PD, Koehler RC, Blizzard KK, Traystman RJ. Deferoxamine reduces early metabolic failure associated with severe cerebral ischemic acidosis in dogs. Stroke. 1995 Apr;26(4):688-94; discussion 94-5.

[269] Nelson CW, Wei EP, Povlishock JT, Kontos HA, Moskowitz MA. Oxygen radicals in cerebral ischemia. Am J Physiol. 1992 Nov;263(5 Pt 2):H1356-62.

[270] Liachenko S, Tang P, Xu Y. Deferoxamine improves early postresuscitation reperfusion after prolonged cardiac arrest in rats. J Cereb Blood Flow Metab. 2003 May;23(5):574-81.

[271] Nayini NR, White BC, Aust SD, Huang RR, Indrieri RJ, Evans AT, *et al*. Post resuscitation iron delocalization and malondialdehyde production in the brain following prolonged cardiac arrest. J Free Radic Biol Med. 1985;1(2):111-6.

[272] Groenendaal F, Shadid M, McGowan JE, Mishra OP, van Bel F. Effects of deferoxamine, a chelator of free iron, on NA(+), K(+)-ATPase activity of cortical brain cell membrane during early reperfusion after hypoxia-ischemia in newborn lambs. Pediatr Res. 2000 Oct;48(4):560-4.

[273] Wu H, Wu T, Xu X, Wang J. Iron toxicity in mice with collagenase-induced intracerebral hemorrhage. J Cereb Blood Flow Metab. 2011 May;31(5):1243-50.

[274] Pawate S, Shen Q, Fan F, Bhat NR. Redox regulation of glial inflammatory response to lipopolysaccharide and interferongamma. J Neurosci Res. 2004 Aug 15;77(4):540-51.

[275] Peeters C, van Bel F. Pharmacotherapeutical reduction of post-hypoxic-ischemic brain injury in the newborn. Biol Neonate. 2001;79(3-4):274-80.

[276] Sorond FA, Ratan RR. Ironing-Out Mechanisms of Neuronal Injury under Hypoxic-Ischemic Conditions and Potential Role of Iron Chelators as Neuroprotective Agents. Antioxidants & Redox Signaling. 2000 Fal;2(3):421-36.

[277] Fredenburg AM, Sethi RK, Allen DD, Yokel RA. The pharmacokinetics and blood-brain barrier permeation of the chelators 1,2 dimethly-, 1,2 diethyl-, and 1-[ethan-1'ol]-2-methyl-3-hydroxypyridin-4-one in the rat. Toxicology. 1996;108(3):191-9.

[278] Mitchell KM, Dotson AL, Cool KM, Chakrabarty A, Benedict SH, LeVine SM. Deferiprone, an orally deliverable iron chelator, ameliorates experimental autoimmune encephalomyelitis. Mult Scler. 2007 Nov;13(9):1118-26.

[279] Song D, Song Y, Hadziahmetovic M, Zhong Y, Dunaief JL. Systemic administration of the iron chelator deferiprone protects against light-induced photoreceptor degeneration in the mouse retina. Free Radic Biol Med. 2012 Jul 1;53(1):64-71.

[280] Bergeron RJ, Liu ZR, McManis JS, Wiegand J. Structural alterations in desferrioxamine compatible with iron clearance in animals. J Med Chem. 1992 Dec 11;35(25):4739-44.

[281] Bergeron RJ, Wiegand J, Brittenham GM. HBED ligand: preclinical studies of a potential alternative to deferoxamine for treatment of chronic iron overload and acute iron poisoning. Blood. 2002 Apr 15;99(8):3019-26.

[282] Kalivendi SV, Kotamraju S, Cunningham S, Shang T, Hillard CJ, Kalyanaraman B. 1-Methyl-4-phenylpyridinium (MPP+)-induced apoptosis and mitochondrial oxidant generation: role of transferrin-receptor-dependent iron and hydrogen peroxide. Biochem J. 2003 Apr 1;371(Pt 1):151-64.

[283] Liang LP, Jarrett SG, Patel M. Chelation of mitochondrial iron prevents seizure-induced mitochondrial dysfunction and neuronal injury. J Neurosci. 2008 Nov 5;28(45):11550-6.

[284] Bar-Am O, Yogev-Falach M, Amit T, Sagi Y, Youdim MB. Regulation of protein kinase C by the anti-Parkinson drug, MAO-B inhibitor, rasagiline and its derivatives, *in vivo*. J Neurochem. 2004 Jun;89(5):1119-25.

[285] Bar Am O, Amit T, Youdim MB. Contrasting neuroprotective and neurotoxic actions of respective metabolites of anti-Parkinson drugs rasagiline and selegiline. Neurosci Lett. 2004 Jan 30;355(3):169-72.

[286] Weinreb O, Bar-Am O, Amit T, Chillag-Talmor O, Youdim MB. Neuroprotection *via* pro-survival protein kinase C isoforms associated with Bcl-2 family members. FASEB J. 2004 Sep;18(12):1471-3.

[287]  Shachar DB, Kahana N, Kampel V, Warshawsky A, Youdim MB. Neuroprotection by a novel brain permeable iron chelator, VK-28, against 6-hydroxydopamine lession in rats. Neuropharmacology. 2004 Feb;46(2):254-63.

[288]  Gal S, Zheng H, Fridkin M, Youdim MB. Novel multifunctional neuroprotective iron chelator-monoamine oxidase inhibitor drugs for neurodegenerative diseases. *In vivo* selective brain monoamine oxidase inhibition and prevention of MPTP-induced striatal dopamine depletion. J Neurochem. 2005 Oct;95(1):79-88.

[289]  Lortie C, King GM, Adamson IY. Effects of dexamethasone on macrophages in fetal and neonatal rat lung. Pediatr Pulmonol. 1990;8(3):138-44.

[290]  Ling EA. Influence of cortisone on amoeboid microglia and microglial cells in the corpus callosum in postnatal rats. J Anat. 1982 Jun;134(Pt 4):705-17.

[291]  Kaur C, Wu CH, Wen CY, Ling EA. The effects of subcutaneous injections of glucocorticoids on amoeboid microglia in postnatal rats. Arch Histol Cytol. 1994 Dec;57(5):449-59.

[292]  Ma W, Gee K, Lim W, Chambers K, Angel JB, Kumar A. Dexamethasone inhibits IL-12p40 production in lipopolysaccharide-stimulated human monocytic cells by down-regulating the activity of c-jun N-terminal kinase, the activation protein-1, and NF-kappa B transcription factors. J Immunol. 2004 Jan 1;172(1):318-30.

[293]  Blotta MH, DeKruyff RH, Umetsu DT. Corticosteroids inhibit IL-12 production in human monocytes and enhance their capacity to induce IL-4 synthesis in CD4(+) lymphocytes. J Immunol. 1997 Jun 15;158(12):5589-95.

[294]  Wilckens T, De Rijk R. Glucocorticoids and immune function: unknown dimensions and new frontiers. Immunol Today. 1997 Sep;18(9):418-24.

[295]  Huo Y, Rangarajan P, Ling EA, Dheen ST. Dexamethasone inhibits the Nox-dependent ROS production *via* suppression of MKP-1-dependent MAPK pathways in activated microglia. BMC Neurosci. 2011;12:49.

[296]  Zhou Y, Ling EA, Dheen ST. Dexamethasone suppresses monocyte chemoattractant protein-1 production *via* mitogen activated protein kinase phosphatase-1 dependent inhibition of Jun N-terminal kinase and p38 mitogen-activated protein kinase in activated rat microglia. J Neurochem. 2007 Aug;102(3):667-78.

[297]  Zhu S, Stavrovskaya IG, Drozda M, Kim BY, Ona V, Li M, *et al*. Minocycline inhibits cytochrome c release and delays progression of amyotrophic lateral sclerosis in mice. Nature. 2002 May 2;417(6884):74-8.

[298]  Thomas M, Le WD. Minocycline: neuroprotective mechanisms in Parkinson's disease. Curr Pharm Des. 2004;10(6):679-86.

[299]  Popovic N, Schubart A, Goetz BD, Zhang SC, Linington C, Duncan ID. Inhibition of autoimmune encephalomyelitis by a tetracycline. Ann Neurol. 2002 Feb;51(2):215-23.

[300]  Teng YD, Choi H, Onario RC, Zhu S, Desilets FC, Lan S, *et al*. Minocycline inhibits contusion-triggered mitochondrial cytochrome c release and mitigates functional deficits after spinal cord injury. Proc Natl Acad Sci U S A. 2004 Mar 2;101(9):3071-6.

[301]  Cho KO, La HO, Cho YJ, Sung KW, Kim SY. Minocycline attenuates white matter damage in a rat model of chronic cerebral hypoperfusion. J Neurosci Res. 2006 Feb 1;83(2):285-91.

[302]  Suk K. Minocycline suppresses hypoxic activation of rodent microglia in culture. Neurosci Lett. 2004 Aug 12;366(2):167-71.

[303]  Tikka TM, Koistinaho JE. Minocycline provides neuroprotection against N-methyl-D-aspartate neurotoxicity by inhibiting microglia. J Immunol. 2001 Jun 15;166(12):7527-33.

[304]  Lechpammer M, Manning SM, Samonte F, Nelligan J, Sabo E, Talos DM, *et al*. Minocycline treatment following hypoxic/ischaemic injury attenuates white matter injury in a rodent model of periventricular leucomalacia. Neuropathol Appl Neurobiol. 2008 Aug;34(4):379-93.

[305]  Zhao F, Hua Y, He YD, Keep RF, Xi GH. Minocycline-Induced Attenuation of Iron Overload and Brain Injury After Experimental Intracerebral Hemorrhage. Stroke. 2011 Dec;42(12):3587-93.

[306]  Potvin F, Petitclerc E, Marceau F, Poubelle PE. Mechanisms of action of antimalarials in inflammation-Induction of apoptosis in human endothelial cells. J Immunol. 1997 Feb 15;158(4):1872-9.

[307]  Karres I, Kremer JP, Dietl I, Steckholzer U, Jochum M, Ertel W. Chloroquine inhibits pro-inflammatory cytokine release into human whole blood. American Journal of Physiology-Regulatory Integrative and Comparative Physiology. 1998 Apr;274(4):R1058-R64.

[308]  Park J, Kwon D, Choi C, Oh JW, Benveniste EN. Chloroquine induces activation of nuclear factor-kappaB and subsequent expression of pro-inflammatory cytokines by human astroglial cells. J Neurochem. 2003 Mar;84(6):1266-74.

[309]  Kaur C, Wu CH, Singh J, Ling EA. Response of amoeboid microglial cells to chloroquine injections in postnatal rats. J Hirnforsch. 1996;37(2):233-42.

[310]  Chua T, Tran T, Yang F, Beech W, Cole GM, Frautschy SA. Effect of chloroquine and leupeptin on intracellular accumulation of amyloid-beta (A beta) 1-42 peptide in a murine N9 microglial cell line. FEBS Lett. 1998 Oct 9;436(3):439-44.

[311]  Lim GP, Yang F, Chu T, Chen P, Beech W, Teter B, *et al*. Ibuprofen suppresses plaque pathology and inflammation in a mouse model for Alzheimer's disease. J Neurosci. 2000 Aug 1;20(15):5709-14.

[312]  Jin DQ, Sung JY, Hwang YK, Kwon KJ, Han SH, Min SS, *et al*. Dexibuprofen (S(+)-isomer ibuprofen) reduces microglial activation and impairments of spatial working memory induced by chronic lipopolysaccharide infusion. Pharmacol Biochem Behav. 2008 May;89(3):404-11.

[313]  Carty ML, Wixey JA, Reinebrant HE, Gobe G, Colditz PB, Buller KM. Ibuprofen inhibits neuroinflammation and attenuates white matter damage following hypoxia-ischemia in the immature rodent brain. Brain Res. 2011 Jul 21;1402:9-19.

[314]  Leroux P, Hennebert C, Catteau J, Legros H, Hennebert O, Laudenbach V, *et al*. Neuroprotective Effects Vary across Nonsteroidal Antiinflammatory Drugs in a Mouse Model of Developing Excitotoxic Brain Injury. Neuroscience. 2010 May 19;167(3):716-23.

[315]  Pandi-Perumal SRS, V; Maestroni, G J M; Cardinali, D P; Poeggeler, B; Hardeland, R. Melatonin: Nature's most versatile biological signal? The FEBS journal. 2006;273(13):2813-38.

[316]  Maestroni GJM. The immunotherapeutic potential of melatonin. Expert Opin Investig Drugs. 2001 Mar;10(3):467-76.

[317]  Hutton LC, Abbass M, Dickinson H, Ireland Z, Walker DW. Neuroprotective properties of melatonin in a model of birth asphyxia in the spiny mouse (Acomys cahirinus). Dev Neurosci. 2009;31(5):437-51.

[318]  Hamada F, Watanabe K, Wakatsuki A, Nagai R, Shinohara K, Hayashi Y, *et al*. Therapeutic Effects of Maternal Melatonin Administration on Ischemia/Reperfusion-Induced Oxidative Cerebral Damage in Neonatal Rats. Neonatology. 2010;98(1):33-40.

[319] Kaur C, Sivakumar V, Ling EA. Melatonin protects periventricular white matter from damage due to hypoxia. J Pineal Res. 2010 Apr;48(3):185-93.

[320] Degos V, Loron G, Mantz J, Gressens P. Neuroprotective strategies for the neonatal brain. Anesth Analg. 2008 Jun;106(6):1670-80.

[321] Kaur C, Ling EA. Effects of melatonin on macrophages/microglia in postnatal rat brain. J Pineal Res. 1999 Apr;26(3):158-68.

[322] Min KJ, Jang JH, Kwon TK. Inhibitory effects of melatonin on the lipopolysaccharide-induced CC chemokine expression in BV2 murine microglial cells are mediated by suppression of Akt-induced NF-kappaB and STAT/GAS activity. J Pineal Res. 2012 Apr;52(3):296-304.

[323] Tocharus J, Chongthammakun S, Govitrapong P. Melatonin inhibits amphetamine-induced nitric oxide synthase mRNA overexpression in microglial cell lines. Neurosci Lett. 2008 Jul 11;439(2):134-7.

[324] Zhou J, Zhang S, Zhao X, Wei T. Melatonin impairs NADPH oxidase assembly and decreases superoxide anion production in microglia exposed to amyloid-beta1-42. J Pineal Res. 2008 Sep;45(2):157-65.

[325] Kubo K, Nakao S, Jomura S, Sakamoto S, Miyamoto E, Xu Y, *et al*. Edaravone, a free radical scavenger, mitigates both gray and white matter damages after global cerebral ischemia in rats. Brain Res. 2009 Jul 7;1279:139-46.

[326] Watanabe T, Yuki S, Egawa M, Nishi H. Protective effects of MCI-186 on cerebral ischemia: possible involvement of free radical scavenging and antioxidant actions. J Pharmacol Exp Ther. 1994 Mar;268(3):1597-604.

[327] Nishi H, Watanabe T, Sakurai H, Yuki S, Ishibashi A. Effect of MCI-186 on brain edema in rats. Stroke. 1989 Sep;20(9):1236-40.

[328] Abe K, Yuki S, Kogure K. Strong attenuation of ischemic and postischemic brain edema in rats by a novel free radical scavenger. Stroke. 1988 Apr;19(4):480-5.

[329] Zhang W, Sato K, Hayashi T, Omori N, Nagano I, Kato S, *et al*. Extension of ischemic therapeutic time window by a free radical scavenger, Edaravone, reperfused with tPA in rat brain. Neurol Res. 2004 Apr;26(3):342-8.

[330] Banno M, Mizuno T, Kato H, Zhang G, Kawanokuchi J, Wang J, *et al*. The radical scavenger edaravone prevents oxidative neurotoxicity induced by peroxynitrite and activated microglia. Neuropharmacology. 2005 Feb;48(2):283-90.

[331] Muller A, Cadenas E, Graf P, Sies H. A novel biologically active seleno-organic compound--I. Glutathione peroxidase-like activity *in vitro* and antioxidant capacity of PZ 51 (Ebselen). Biochem Pharmacol. 1984 Oct 15;33(20):3235-9.

[332] Schewe C, Schewe T, Wendel A. Strong inhibition of mammalian lipoxygenases by the antiinflammatory seleno-organic compound ebselen in the absence of glutathione. Biochem Pharmacol. 1994 Jul 5;48(1):65-74.

[333] Hattori R, Inoue R, Sase K, Eizawa H, Kosuga K, Aoyama T, *et al*. Preferential inhibition of inducible nitric oxide synthase by ebselen. Eur J Pharmacol. 1994 Apr 15;267(2):R1-2.

[334] Cotgreave IA, Duddy SK, Kass GE, Thompson D, Moldeus P. Studies on the anti-inflammatory activity of ebselen. Ebselen interferes with granulocyte oxidative burst by dual inhibition of NADPH oxidase and protein kinase C? Biochem Pharmacol. 1989 Feb 15;38(4):649-56.

[335] Patrick RA, Peters PA, Issekutz AC. Ebselen is a specific inhibitor of LTB4-mediated migration of human neutrophils. Agents Actions. 1993;40(3-4):186-90.

[336]  Shimohashi N, Nakamuta M, Uchimura K, Sugimoto R, Iwamoto H, Enjoji M, *et al.* Selenoorganic compound, ebselen, inhibits nitric oxide and tumor necrosis factor-alpha production by the modulation of jun-N-terminal kinase and the NF-kappab signaling pathway in rat Kupffer cells. J Cell Biochem. 2000 Jun 12;78(4):595-606.

[337]  Banati RB, Gehrmann J, Schubert P, Kreutzberg GW. Cytotoxicity of microglia. Glia. 1993 Jan;7(1):111-8.

[338]  Takasago T, Peters EE, Graham DI, Masayasu H, Macrae IM. Neuroprotective efficacy of ebselen, an antioxidant with anti-inflammatory actions, in a rodent model of permanent middle cerebral artery occlusion. Br J Pharmacol. 1997 Nov;122(6):1251-6.

[339]  Imai H, Masayasu H, Dewar D, Graham DI, Macrae IM. Ebselen protects both gray model of focal and white matter in a rodent cerebral ischemia. Stroke. 2001 Sep;32(9):2149-56.

[340]  Ralat LA, Colman RF. Identification of tyrosine 79 in the tocopherol binding site of glutathione S-transferase Pi. Biochemistry (Mosc). 2006 Oct 17;45(41):12491-9.

[341]  Li YK, Liu L, Barger SW, Mrak RE, Griffin WST. Vitamin E suppression of microglial activation is neuroprotective. J Neurosci Res. 2001 Oct 15;66(2):163-70.

[342]  Egger T, Schuligoi R, Wintersperger A, Amann R, Malle E, Sattler W. Vitamin E (alpha-tocopherol) attenuates cyclo-oxygenase 2 transcription and synthesis in immortalized murine BV-2 microglia. Biochem J. 2003 Mar 1;370:459-67.

[343]  Heppner FL, Roth K, Nitsch R, Hailer NP. Vitamin E induces ramification and downregulation of adhesion molecules in cultured microglial cells. Glia. 1998 Feb;22(2):180-8.

[344]  Sporn MB, Roberts AB. Cervical Dysplasia Regression Induced by All-Trans-Retinoic Acid. J Natl Cancer Inst. 1994 Apr 6;86(7):476-7.

[345]  Datta PK, Reddy RS, Lianos EA. Effects of all-trans-retinoic acid (atRA) on inducible nitric oxide synthase (iNOS) activity and transforming growth factor beta-1 production in experimental anti-GBM antibody-mediated glomerulonephritis. Inflammation. 2001 Dec;25(6):351-9.

[346]  Mehta K, Mcqueen T, Tucker S, Pandita R, Aggarwal BB. Inhibition by All-Trans-Retinoic Acid of Tumor-Necrosis-Factor and Nitric-Oxide Production by Peritoneal-Macrophages. J Leukoc Biol. 1994 Mar;55(3):336-42.

[347]  Heyman RA, Mangelsdorf DJ, Dyck JA, Stein RB, Eichele G, Evans RM, *et al.* 9-Cis Retinoic Acid Is a High-Affinity Ligand for the Retinoid-X Receptor. Cell. 1992 Jan 24;68(2):397-406.

[348]  Dheen ST, Jun Y, Yan Z, Tay SS, Ling EA. Retinoic acid inhibits expression of TNF-alpha and iNOS in activated rat microglia. Glia. 2005 Apr 1;50(1):21-31.

[349]  Eder C. Ion channels in microglia (brain macrophages). American Journal of Physiology-Cell Physiology. 1998 Aug;275(2):C327-C42.

[350]  Skaper SD. Ion Channels on Microglia: Therapeutic Targets for Neuroprotection. Cns & Neurological Disorders-Drug Targets. 2011 Feb;10(1):44-56.

[351]  Koo GC, Blake JT, Talento A, Nguyen M, Lin S, Sirotina A, *et al.* Blockade of the voltage-gated potassium channel Kv1.3 inhibits immune responses *in vivo*. J Immunol. 1997 Jun 1;158(11):5120-8.

[352]  Koo GC, Blake JT, Shah K, Staruch MJ, Dumont F, Wunderler D, *et al.* Correolide and derivatives are novel immunosuppressants blocking the lymphocyte Kv1.3 potassium channels. Cell Immunol. 1999 Nov 1;197(2):99-107.

[353] Beeton C, Barbaria J, Giraud P, Devaux J, Benoliel AM, Gola M, *et al.* Selective blocking of voltage-gated K+ channels improves experimental autoimmune encephalomyelitis and inhibits T cell activation. J Immunol. 2001 Jan 15;166(2):936-44.

[354] Uitdehaag BMJ, Polman CH, Degroot CJA, Dijkstra CD. Effect of K+ Channel Blockers on the Clinical Course and Histological Features of Experimental Allergic Encephalomyelitis. Acta Neurol Scand. 1994 Oct;90(4):299-301.

[355] Li F, Lu J, Wu CY, Kaur C, Sivakumar V, Sun J, *et al.* Expression of Kv1.2 in microglia and its putative roles in modulating production of pro-inflammatory cytokines and reactive oxygen species. J Neurochem. 2008 Sep;106(5):2093-105.

[356] Khanna R, Roy L, Zhu XP, Schlichter LC. K+ channels and the microglial respiratory burst. American Journal of Physiology-Cell Physiology. 2001 Apr;280(4):C796-C806.

[357] Eder C, Klee R, Heinemann U. Pharmacological properties of Ca2+-activated K+ currents of ramified murine brain macrophages. Naunyn-Schmiedebergs Archives of Pharmacology. 1997 Aug;356(2):233-9.

[358] Farber K, Kettenmann H. Functional role of calcium signals for microglial function. Glia. 2006 Nov 15;54(7):656-65.

[359] Moller T, Contos JJ, Musante DB, Chun J, Ransom BR. Expression and function of lysophosphatidic acid receptors in cultured rodent microglial cells. J Biol Chem. 2001 Jul 13;276(28):25946-52.

[360] D'Aversa TG, Yu KOA, Berman JW. Expression of chemokines by human fetal microglia after treatment with the human immunodeficiency virus type 1 protein Tat. J Neurovirol. 2004 Apr;10(2):86-97.

[361] Inoue K, Nakajima K, Morimoto T, Kikuchi Y, Koizumi S, Illes P, *et al.* ATP stimulation of Ca2+-dependent plasminogen release from cultured microglia. Br J Pharmacol. 1998 Apr;123(7):1304-10.

[362] Kohler M, Hirschberg B, Bond CT, Kinzie JM, Marrion NV, Maylie J, *et al.* Small-conductance, calcium-activated potassium channels from mammalian brain. Science. 1996 Sep 20;273(5282):1709-14.

[363] Stocker M. Ca(2+)-activated K(+) channels: Molecular determinants and function of the SK family. Nature Reviews Neuroscience. 2004 Oct;5(10):758-70.

[364] Xia XM, Fakler B, Rivard A, Wayman G, Johnson-Pais T, Keen JE, *et al.* Mechanism of calcium gating in small-conductance calcium-activated potassium channels. Nature. 1998 Oct 1;395(6701):503-7.

[365] Wulff H, Miller MJ, Hansel W, Grissmer S, Cahalan MD, Chandy KG. Design of a potent and selective inhibitor of the intermediate-conductance Ca2+-activated K+ channel, IKCa1: A potential immunosuppressant. Proc Natl Acad Sci U S A. 2000 Jul 5;97(14):8151-6.

[366] Kaushal V, Koeberle PD, Wang Y, Schlichter LC. The Ca2+-activated K+ channel KCNN4/KCa3.1 contributes to microglia activation and nitric oxide-dependent neurodegeneration. J Neurosci. 2007 Jan 3;27(1):234-44.

[367] Xiang ZH, Burnstock G. Expression of P2X receptors on rat microglial cells during early development. Glia. 2005 Nov 1;52(2):119-26.

[368] Li F, Wang L, Li JW, Gong M, He L, Feng R, *et al.* Hypoxia induced amoeboid microglial cell activation in postnatal rat brain is mediated by ATP receptor P2X4. BMC Neurosci. 2011 Nov 4;12.

[369] Domercq M, Perez-Samartin A, Aparicio D, Alberdi E, Pampliega O, Matute C. P2X7 receptors mediate ischemic damage to oligodendrocytes. Glia. 2010 Apr 15;58(6):730-40.

[370]   Haynes SE, Hollopeter G, Yang G, Kurpius D, Dailey ME, Gan WB, *et al.* The P2Y12 receptor regulates microglial activation by extracellular nucleotides. Nat Neurosci. 2006 Dec;9(12):1512-9.

[371]   Gendron FP, Chalimoniuk M, Strosznajder J, Shen S, Gonzalez FA, Weisman GA, *et al.* P2X7 nucleotide receptor activation enhances IFN gamma-induced type II nitric oxide synthase activity in BV-2 microglial cells. J Neurochem. 2003 Oct;87(2):344-52.

[372]   Ohsawa K, Irino Y, Nakamura Y, Akazawa C, Inoue K, Kohsaka S. Involvement of P2X4 and P2Y12 receptors in ATP-induced microglial chemotaxis. Glia. 2007 Apr 15;55(6):604-16.

[373]   Light AR, Wu Y, Hughen RW, Guthrie PB. Purinergic receptors activating rapid intracellular Ca increases in microglia. Neuron Glia Biol. 2006 May;2(2):125-38.

[374]   Ling C, Li Y, Zhu X, Zhang C, Li M. Ginsenosides may reverse the dexamethasone-induced down-regulation of glucocorticoid receptor. Gen Comp Endocrinol. 2005;140(3):203-9.

[375]   Leung KW, Wong AS. Pharmacology of ginsenosides: a literature review. Chin Med. 2010;5:20.

[376]   Wu CF, Bi XL, Yang JY, Zhan JY, Dong YX, Wang JH, *et al.* Differential effects of ginsenosides on NO and TNF-alpha production by LPS-activated N9 microglia. Int Immunopharmacol. 2007 Mar;7(3):313-20.

[377]   Zong Y, Ai QL, Zhong LM, Dai JN, Yang P, He Y, *et al.* Ginsenoside Rg1 attenuates lipopolysaccharide-induced inflammatory responses *via* the phospholipase C-gamma1 signaling pathway in murine BV-2 microglial cells. Curr Med Chem. 2012;19(5):770-9.

[378]   Bae EA, Kim EJ, Park JS, Kim HS, Ryu JH, Kim DH. Ginsenosides Rg3 and Rh2 inhibit the activation      of      AP-1      and      protein      kinase      A      pathway      in lipopolysaccharide/interferon-gamma-stimulated BV-2 microglial cells. Planta Med. 2006 Jun;72(7):627-33.

[379]   Joo SS, Lee DI. Potential effects of microglial activation induced by ginsenoside Rg3 in rat primary culture: enhancement of type A Macrophage Scavenger Receptor expression. Arch Pharm Res. 2005 Oct;28(10):1164-9.

[380]   Xu X, Lu YY, Bie XD. Protective effects of gastrodin on hypoxia-induced toxicity in primary cultures of rat cortical neurons. Planta Med. 2007 Jun;73(7):650-4.

[381]   Shu C, Chen C, Zhang DP, Guo H, Zhou H, Zong J, *et al.* Gastrodin protects against cardiac hypertrophy and fibrosis. Mol Cell Biochem. 2012 Jan;359(1-2):9-16.

[382]   Meng qy. Effects of Tianma injection on transformation of the splenic lymphocyte in mice. Guangxi Zhongyiyao. 2003;26(2):52-4.

[383]   Dai JN, Zong Y, Zhong LM, Li YM, Zhang W, Bian LG, *et al.* Gastrodin Inhibits Expression of Inducible NO Synthase, Cyclooxygenase-2 and Pro-inflammatory Cytokines in Cultured LPS-Stimulated Microglia *via* MAPK Pathways. PLoS ONE. 2011 Jul 12;6(7).

[384]   Huang NK, Lin YL, Cheng JJ, Lai WL. Gastrodia elata prevents rat pheochromocytoma cells from serum-deprived apoptosis: the role of the MAPK family. Life Sci. 2004 Aug 13;75(13):1649-57.

[385]   Kim sT. Neuroprotective Effect of Some Plant Extracts in Cultured CT105-Induced PC12 Cells. Biol Pharm Bull. 2006;29(10):2021-4.

[386]   Choi DK, Koppula S, Suk K. Inhibitors of Microglial Neurotoxicity: Focus on Natural Products. Molecules. 2011 Feb;16(2):1021-43.

[387]   Saiko P, Szakmary A, Jaeger W, Szekeres T. Resveratrol and its analogs: Defense against cancer, coronary disease and neurodegenerative maladies or just a fad? Mutation Research-Reviews in Mutation Research. 2008 Jan-Feb;658(1-2):68-94.

[388]  Das S, Das DK. Anti-inflammatory responses of resveratrol. Inflamm Allergy Drug Targets. 2007 Sep;6(3):168-73.

[389]  Candelario-Jalil E, de Oliveira AC, Graf S, Bhatia HS, Hull M, Munoz E, *et al.* Resveratrol potently reduces prostaglandin E2 production and free radical formation in lipopolysaccharide-activated primary rat microglia. J Neuroinflammation. 2007;4:25.

[390]  Bi XL, Yang JY, Dong YX, Wang JM, Cui YH, Ikeshima T, *et al.* Resveratrol inhibits nitric oxide and TNF-alpha production by lipopolysaccharide-activated microglia. Int Immunopharmacol. 2005 Jan;5(1):185-93.

[391]  Zhang Z, Yang Y, Pang W, Sun C, Yang G. [Effect and underlying mechanism of resveratol on porcine primary preadipocyte apoptosis]. Sheng Wu Gong Cheng Xue Bao. 2010 Aug;26(8):1042-9.

[392]  Karalis F, Soubasi V, Georgiou T, Nakas CT, Simeonidou C, Guiba-Tziampiri O, *et al.* Resveratrol ameliorates hypoxia/ischemia-induced behavioral deficits and brain injury in the neonatal rat brain. Brain Res. 2011 Nov 24;1425:98-110.

[393]  Lee HS, Jung KK, Cho JY, Rhee MH, Hong S, Kwon M, *et al.* Neuroprotective effect of curcumin is mainly mediated by blockade of microglial cell activation. Pharmazie. 2007 Dec;62(12):937-42.

[394]  Kang G, Kong PJ, Yuh YJ, Lim SY, Yim SV, Chun W, *et al.* Curcumin suppresses lipopolysaccharide-induced cyclooxygenase-2 expression by inhibiting activator protein 1 and nuclear factor kappab bindings in BV2 microglial cells. J Pharmacol Sci. 2004 Mar;94(3):325-8.

[395]  Kim HYP, Eun Jung; Joe, Eun-Hye; Jou, Ilo. Curcumin suppresses Janus kinase-STAT inflammatory signaling through activation of Src homology 2 domain-containing tyrosine phosphatase 2 in brain microglia. Journal of immunology (Baltimore, Md : 1950). 2003;171(11):6072-9.

[396]  Karlstetter M, Lippe E, Walczak Y, Moehle C, Aslanidis A, Mirza M, *et al.* Curcumin is a potent modulator of microglial gene expression and migration. J Neuroinflammation. 2011 Sep 29;8.

[397]  He LF, Chen HJ, Qian LH, Chen GY, Buzby JS. Curcumin protects pre-oligodendrocytes from activated microglia *in vitro* and *in vivo*. Brain Res. 2010 Jun 21;1339:60-9.

[398]  Lyss G, Knorre A, Schmidt TJ, Pahl HL, Merfort I. The anti-inflammatory sesquiterpene lactone helenalin inhibits the transcription factor NF-kappaB by directly targeting p65. J Biol Chem. 1998 Dec 11;273(50):33508-16.

[399]  Eliza J, Daisy P, Ignacimuthu S. Antioxidant activity of costunolide and eremanthin isolated from Costus speciosus (Koen ex. Retz) Sm. Chem Biol Interact. 2010 Dec 5;188(3):467-72.

[400]  Xu J, Jin D-Q, Zhao P, Song X, Sun Z, Guo Y, *et al.* Sesquiterpenes inhibiting NO production from Celastrus orbiculatus. Fitoterapia. 2012.

[401]  Suuronen T, Huuskonen J, Pihlaja R, Kyrylenko S, Salminen A. Regulation of microglial inflammatory response by histone deacetylase inhibitors. J Neurochem. 2003 Oct;87(2):407-16.

[402]  Rayan NA, Baby N, Pitchai D, Indraswari F, Ling EA, Lu J, *et al.* Costunolide inhibits pro-inflammatory cytokines and iNOS in activated murine BV2 microglia. Front Biosci (Elite Ed). 2011;3:1079-91.

[403]  Fiebich BL, Grozdeva M, Hess S, Hull M, Danesch U, Bodensieck A, *et al.* Petasites hybridus extracts *in vitro* inhibit COX-2 and PGE2 release by direct interaction with the

enzyme and by preventing p42/44 MAP kinase activation in rat primary microglial cells. Planta Med. 2005 Jan;71(1):12-9.

[404] Bonfoco E, Ceccatelli S, Manzo L, Nicotera P. Colchicine Induces Apoptosis in Cerebellar Granule Cells. Exp Cell Res. 1995 May;218(1):189-200.

[405] Tsukidate K, Yamamoto K, Synder JW, Farber JL. Microtubule Antagonists Activate Programmed Cell-Death (Apoptosis) in Cultured Rat Hepatocytes. Am J Pathol. 1993 Sep;143(3):918-25.

[406] Kaur C. Effects of colchicine on amoeboid microglial cells in the postnatal rat brain. Arch Histol Cytol. 1997 Dec;60(5):453-62.

[407] Dommergues MA, Plaisant F, Verney C, Gressens P. Early microglial activation following neonatal excitotoxic brain damage in mice: A potential target for neuroprotection. Neuroscience. 2003;121(3):619-28.

[408] Chen Y, Liu L. Modern methods for delivery of drugs across the blood-brain barrier. Adv Drug Deliv Rev. 2012 May 15;64(7):640-65.

[409] Saunders NR, Dziegielewska KM, Mollgard K. The importance of the blood-brain barrier in fetuses and embryos. Trends Neurosci. 1991 Jan;14(1):14-5.

[410] Mollgard K, Saunders NR. The development of the human blood-brain and blood-CSF barriers. Neuropathol Appl Neurobiol. 1986 Jul-Aug;12(4):337-58.

[411] deVries HE, BlomRoosemalen MCM, vanOosten M, deBoer AG, vanBerkel TJC, Breimer DD, *et al.* The influence of cytokines on the integrity of the blood-brain barrier *in vitro*. J Neuroimmunol. 1996 Jan;64(1):37-43.

[412] Abbott NJ. Inflammatory mediators and modulation of blood-brain barrier permeability. Cell Mol Neurobiol. 2000 Apr;20(2):131-47.

[413] Deli MA. Potential use of tight junction modulators to reversibly open membranous barriers and improve drug delivery. Bba-Biomembranes. 2009 Apr;1788(4):892-910.

[414] Hynynen K. Ultrasound for drug and gene delivery to the brain. Adv Drug Deliver Rev. 2008 Jun 30;60(10):1209-17.

[415] Stam R. Electromagnetic fields and the blood-brain barrier. Brain research reviews. 2010 Oct 5;65(1):80-97.

[416] Dai H, Navath RS, Balakrishnan B, Guru BR, Mishra MK, Romero R, *et al.* Intrinsic targeting of inflammatory cells in the brain by polyamidoamine dendrimers upon subarachnoid administration. Nanomed. 2010 Nov;5(9):1317-29.

[417] Kannan RM, Kannan S, Romero R, *et al.* US10/38068. Dendrimer-based therapeutic nanodevices for therapeutic and imaging applications, **2010**.

[418] Kannan RM, Iezzi R, Kannan S, Guru BR. Dendrimer-containing particles for sustained release of compounds. US provisional patent filed May 2007, (application 60/997987). International patent filed October 2008 (application PCT/US2008/078988), Regular patents filed, 2010.

# INDEX

## A

**Atta-ur-Rahman (Ed)**
**All rights reserved-© 2013 Bentham Science Publishers**

# G

# H

www.ingramcontent.com/pod-product-compliance
Lightning Source LLC
Chambersburg PA
CBHW050824220326
41598CB00006B/308